This book gives an account of the evolution of our major crop plants, of their migration with man from their centres of origin and of the consequent development of an astonishing range of locally adapted landraces. The wild ancestral species, the landraces and the early varieties derived from them by breeding constitute the genetic resources of each crop. The book traces the impact of purposive breeding and selection, and of the substitution of uniformity for diversity, thereby destroying the resource base on which further change depends.

The collecting of crop genetic diversity, its conservation and management for future use are discussed. The role of conserved genetic diversity in the current movement to find low-input sustainable systems of food production is considered. Throughout the book, the dependence of adaptive change on the availability of genetic diversity is stressed as the key to survival in a changing world. The book considers the need for world-wide collaboration in conservation, outlines gaps in the science base, and suggests some urgent research needs. It concludes by placing crop plant genetic diversity in the wider context of the conservation of the biosphere.

This book is intended as a non-technical introduction to the conservation of crop genetic diversity for members of the public, policy-makers, and undergraduates and postgraduates in agricultural, biological and environmental sciences.

T0282310

GENES, CROPS AND THE ENVIRONMENT

GENES, CROPS AND THE ENVIRONMENT

JOHN HOLDEN, JAMES PEACOCK AND
TREVOR WILLIAMS

CAMBRIDGE
UNIVERSITY PRESS

CAMBRIDGE UNIVERSITY PRESS
Cambridge, New York, Melbourne, Madrid, Cape Town, Singapore,
São Paulo, Delhi, Dubai, Tokyo, Mexico City

Cambridge University Press
The Edinburgh Building, Cambridge CB2 8RU, UK

Published in the United States of America by Cambridge University Press, New York

www.cambridge.org
Information on this title: www.cambridge.org/9780521437370

First published 1993

A catalogue record for this publication is available from the British Library

Library of Congress cataloguing in publication data
Holden. J. H. W. (John H. W.)
Genes. crops, and the environment / John Holden, James Peacock, and Trevor Williams.
p. cm.
Includes bibliographical references and index.
ISBN 0-521-43137-9. - ISBN 0-521-43737-7 (pbk.)
I. Crops-Germplasm resources. 2. Germplasm resources, Plant.
I. Peacock, W. J. (William James) II. Williams, John Trevor.
Ill. Title.
SB123.3.H65 1993
333.95'316—dc20 92-33605 CIP

ISBN 978-0-521-43137-8 Hardback
ISBN 978-0-521-43737-0 Paperback

Contents

Preface

On hearing that this book was in preparation, a scientist friend remarked, 'Surely not another book on plant genetic resources'. Friends, of course, have the right to make candid remarks, and scientific friends seem to exercise the right freely. In this case the comment was, we think, inappropriate.

It is true that a considerable literature has built up on this subject in the past fifteen years, in the form of books, papers in scientific journals, periodic newsletters, many specialist technical and scientific reviews, reports and booklets and at least two journals devoted to plant genetic resources issues. These different types of publications, apart from their subject, have one other feature in common; they are addressed largely or wholly to those already working in plant genetic resources conservation.

On the other hand, there has been a steadily increasing output of material in newspapers and popular-science journals, on radio and on television, addressed to the public at large. This publicity has aroused some interest in an issue which, despite its implications for us all, might otherwise have remained as another obscure branch of applied science and technology, unknown except to those on the inside. However, the mass media presentations have usually dealt with separate aspects of what is a large and complex issue, and inevitably the picture presented has been fragmentary.

Much of the success of these presentations has been due to the intrinsic interest and obvious importance of the subject, but much is also due to the skill of professional communicators, in presenting new ideas in an attractive and accessible form. Scientists too are communicators, but usually they write for their colleagues in the same field, which makes their task easier in that they know that their intended audience is familiar with the origins and meaning of the terminology which they use, but which prevents their ideas being accessible to a wider audience.

In this book, we attempt to unite these two different approaches; to present an overall picture of plant genetic resources conservation – why it is necessary, what it involves and what are the benefits which it is likely to bring to world food production in a time of rapid environmental, social, economic and scientific/technological change – and to reach a wider audience than has been reached by most previous comprehensive treatments.

To this end we have tried to concentrate on essentials and to avoid confusing detail, to illustrate principles and practices and to make as few assumptions as possible about the background knowledge in biological science of our readers. We have avoided the use of technical terms and, where this has not been possible, we have attempted to provide explanations in plain words.

We hope that this book will be of value to scientists of other disciplines, to students, to politicians and policy-makers, but above all to taxpayers who are interested in conservation issues and who ultimately are the ones who determine what resources are devoted to the protection of the environment and the genetic diversity of our crops.

The reference list is not intended to be comprehensive but has been selected to lead the reader, either to key source books or to reviews, which through their contents or bibliographies can provide doorways into wider areas of reading.

It is a pleasure to acknowledge our debt to IBPGR for permission to reproduce Figure 4.1, 6.1 and 6.2 and in particular to Dr Mark Perry for generously providing and allowing us to publish the data in Table 4.1 which, we believe, is the most accurate and up-to-date available on the number of world-wide accessions of crop germplasm. Other sources from which illustrations have been specially prepared, are acknowledged in the figures and tables.

Especial thanks are due to professor Don Marshall who read through the text in draft and made valuable critical comments, to Dr Peter Holden for his expert assistance in the presentation of the figures and tables and, last but not least, to Bronwen Holden, who first drew our attention to the need for a book addressed to the interested layman. We hope that we have not disappointed her expectations, nor those of our other readers.

John Holden

Abbreviations

AVRDC	Asian Vegetable Research and Development Center
CGIAR	Consultative Group on International Agricultural Research
CIAT	Centro Internacional de Agricultura Tropical
CIMMYT	Centro Internacional de Mejoramento de Maiz y Trigo
CIP	Centro Internacional de la Papa
CPC	Commonwealth Potato Collection
DNA	Deoxyribose nucleic acid
FAO	Food and Agriculture Organisation of the United Nations
GRAAS	Genetic Resources Accession Assessment Score
GREWS	Genetic Resources Early Warning Systems
IBP	International Biological Programme
IBPGR	International Board for Plant Genetic Resources
ILDIS	International Legume Database and Information Service
IRRI	International Rice Research Institute
IUCN	World Conservation Union (formerly International Union for the Conservation of Nature and Natural Resources)
IVAG	*In-vitro* Active Genebank
IVBG	*In-vitro* Base Genebank
UNDP	United Nations Development Programme
UNEP	United Nations Environment Programme
UNESCO	United Nations Educational, Scientific and Cultural Organisation
WWF	World Wide Fund for Nature

1

Man and plants: a relationship in crisis

Résumé

Each one of the astounding number of diverse species which make up the flora of the planet, is adapted to fill a niche in the apparently endless range of earth's habitats. Within each species there is a finer degree of adaptation of populations to smaller differences between essentially similar habitats. These adapted populations within species are known as ecotypes. They, and the associated assembly of ecotypes of other species which live in the same habitat, constitute an ecosystem. Ecosystems are subject to change over time. Change within ecosystems is a normal event in the natural world. Its main causes are changes in the climate and other components of the environment. Plants have the means to respond to slow natural changes through the production in every generation of individuals with genes in new combinations and hence with different ecological preferences. The source of new genetic variation is mutation – permanent, heritable changes in genes and their function – and the mechanism for creating new combinations is sexual reproduction.

If species are able to adapt, they survive; if not they perish, and this has been the fate of innumerable species in the past. Yet for many others their responsiveness to environmental change has been adequate, at least until recently. For the past two to three thousand years the rate of environmental change has progressively increased due to the human population explosion and to the increasing exploitation of the natural world through industrialisation and agricultural development. In many areas, plant populations are no longer able to cope with the rate of change and are disappearing. Humans, for long ignorant or uncaring about their destructive exploitation of the environment, have recently begun to show concern about the consequences of past and present follies, because of the dawning realisation of their threat to the biosphere, in general, and to people, in particular. There is a growing awareness of the need for remedial action, which is expressed as conservation activities of different kinds, and most recently as the first attempts to plan strategies for coupling development to the sustainable use of natural resources.

The purpose of this book is to consider in detail one part of the conservation movement, namely the conservation of the genetic diversity of crop plants, a resource which is essential in the adaptation of crops through plant breeding to meet the changing needs of future generations.

Wild plants and plant communities are constantly changing

The seemingly endless variety of living things, plants and animals and micro-organisms, live together in characteristic communities of particular species – ecosystems – and not in random mixtures of species which occupy the same area by chance. The component species of an ecosystem are different in different habitats. The species interact, competitively, with each other and react to their physical environment – the temperature, rainfall, daylength, light intensity and soil type of the habitat – to form a dynamic community of immense complexity which is adapted to survival in that habitat. Ecosystems evolve through successive stages towards a final climax association of species which is relatively stable. However, even the climax community is in a state of dynamic equilibrium, and is liable to change in the long term.

There are two primary causes of instability in ecosystems; changes in climate and changes in human activities. Instability of climate due to long-term directional weather changes is seen in dramatic form in the advance and retreat of the polar ice sheets. The systematic accumulation of weather data in the past fifty to one hundred years and the recent development of methods to estimate the climate of past ages, have begun to quantify what previously had been inferred from anecdotal evidence, namely that climatic changes are normal, that they occur in short and long-term cycles, and that they differ in degree and extent across the globe.

Plant communities have responded to these environmental cycles. The response has occurred with change at three levels: the succession of one ecosystem by another, the substitution of one species for another within an ecosystem and changes in the genetic constitution of the individuals within a species. Generally, change will occur in all of these ways, resulting in the extinction of some species, the emergence of new ones, and in changes in the species composition of communities as well as in their geographical distributions.

We can read the sequences of past vegetation changes from the pollen profiles present in peat bogs and in alluvial sediments. Remains of larger fossils have also enabled scientists to paint the moving picture of vegetation change of past times. This provides a broad-brush picture of successional changes in plant communities. Recent scientific developments in the isolation and analysis of DNA from preserved plant remains indicate the future possibility of comparing the detailed genetic composition of modern with ancient ancestral plants.

The second cause of change, human activity, has the potential for inducing changes which are sudden and irreversible. Large-scale deforestation to provide land for agriculture and wood for fuel, intensive grazing by flocks of domesticated animals, which prevents the regeneration of forest and shrub communities through the destruction of their seedlings, and industrial development and urbanisation, have all had dramatic and directly destructive effects on natural ecosystems. In addition, man has caused equally profound indirect effects through environmental pollution from industrial and urban development. Some of these can rapidly change or destroy communities, such as the general pollution by heavy metals of the water, soil and atmosphere around the sites of heavy manufacturing industry, but others are insidious, cumulative and slower to appear. A notable example of the latter is the much publicised 'greenhouse effect', in which rising levels of atmospheric carbon dioxide, methane and other gases are thought to be the cause of a progressive rise in atmospheric temperature. The full effects of this temperature rise on climate, sea levels, ocean currents and of course on vegetation, have yet to be determined.

It has always been a feature of life on this planet that plant and animal species have had to adapt to changing environments and survival has depended on their ability to do so. The new factor in this dynamic relationship between plants and their environment is the greatly accelerated rate of change imposed by man, with which the response mechanisms of plants are unable to cope, resulting in the large-scale losses, not only of species but also of whole ecosystems.

Genetic variation and sex are the keys to adaptive change

The characteristics of plants are determined by the genes located in their DNA. The expression of many of these genes is subject to the influence of the environment, and so plant characteristics are generally thought of as the result of gene and environment interaction. Genes are themselves subject to constant – though infrequent – changes due to events which are both external (e.g. irradiation) and internal (e.g. errors in the replication and repair of DNA). These random changes in the linear code of the DNA, affect its activity and produce the variation in the form and function of the plant on which adaptation depends. Organisms evolve with new potentialities and attributes which differ in their fitness under the selection pressures of a particular environment.

One attribute which developed early in the evolution of plants and which has since been universally conserved, is the ability to exchange genetic material among individuals. In higher plants, the mechanism of exchange – sexual reproduction – takes many forms, often of great complexity. The widespread occurrence of sexual reproduction in the living world, coupled often with the development of elaborate genetic mechanisms to control the degree of relationship of the mating individuals, is a clear indication of its importance to adaptation, evolution and survival of the species.

Sexual reproduction ensures the spread and mixing of genetic variation within the group of individuals which exchange genes. The consequence is that, in each generation, all individuals differ in their genetic potential and hence their fitness to the environment. Some will be better adapted than others and will leave more progeny. These progeny, in turn, will go through the same process of genetic reassortment when they in turn reproduce.

The normal state of affairs is of plant populations in a state of constant genetic flux, where individuals of different fitness interact with each other and with a physical world, which itself is subject to short and long-term change. If the generation of new forms with suitable attributes is adequate, then survival of the species is possible in the changing world. If it is not then species die out. Clearly, in this interaction between genetic systems and changing environmental pressures, the rate of change of the environment and the rate of genetic response of the species, are critical to survival.

Natural ecosystems are being destroyed by human population increase and industrialisation

Rates of change in the living world are accelerating, due to the progressive disruption of ecosystems by human activity. Adaptive responses, which are necessary for the survival of wild species, are dependent on an adequate reserve of genetic variability. The same is true for the adaptation of crop plants to present and future needs.

However, the possibilities of achieving the necessary adaptive changes are being reduced by the erosion, by human activity, of genetic variability. This is happening directly, through the large-scale destruction of plant communities, and indirectly through man-induced climatic changes lead-

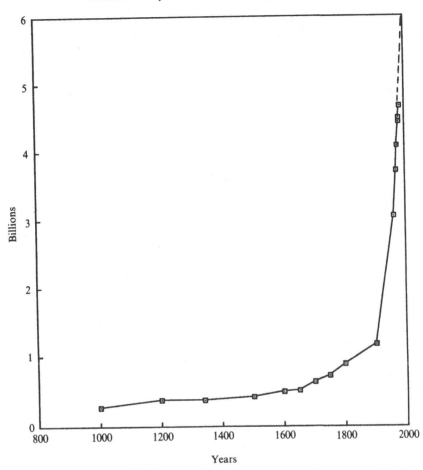

Fig. 1.1 Estimates of world population from AD 1000 to 1983 and extrapolated to the year 2000. (Data from Clark, 1967), and FAO, 1984.)

ing to, for example, the spread of deserts into hitherto semi-arid regions. In agriculture, it is happening at a headlong rate as a consequence of the recent practice of substituting genetically uniform high-yielding varieties, for the ancient and highly variable agro-ecotypes or landraces, on which agriculture has for so long depended.

Man has always been an integral part of the natural ecosystems of the planet, and until recently has lived as a harmonious part of these systems. It is only in the past 200 years, or so, with the deadly combination of runaway increases in population (Fig. 1.1) and poverty and in the ruthless exploitation of the natural resources of the planet, that this harmonious relationship has been destroyed. It has been estimated that when agriculture first developed, the total world population was of the order of ten million (Clark, 1967). Today it is approaching six billion, and increasing rapidly, particularly in the developing countries of Latin America, Africa and Asia.

In the face of pressures for food, water and shelter for these billions of people, of the remorseless demands for raw materials for industry and the associated pollution of the environment, all of our ecosystems are at serious risk.

Public awareness of environmental destruction is growing

Concern about man's degradation of the environment is not new. It has a long and distinguished history which may be traced back in Western thought to the times of classical Greece. For example, fears of widespread man-induced climatic change, often thought to be of recent origin, may be traced back to the writings of Theophrastus, which apparently prompted the forest conservation policies of some British colonial countries (Grove, 1990).

In the 17th and 18th centuries, environmental concerns developed in association with, and as a result of, the colonisation of the world by Europeans and particularly as a result of the growing awareness, among a more thoughtful minority of those involved, of the destructive impact of European development on the environments and peoples of the newly colonised lands. In general, colonisation was regarded as a kind of cultural evangelism, and the exploitation of the natural resources of the subjugate lands was thought of as a natural reward for the civilising benefits which were provided. Rare criticism of development practices came usually from scientists and particularly from ecologists who studied the origins and composition of ecosystems. It goes without saying that these criticisms, published in specialist scientific journals and usually couched in objective, restrained prose, rarely travelled beyond the community of professional biologists and did little to awake public concern to the damage being done or to its implications for the future.

The movement of environmental issues from the specialist concerns of a few scientists to a centre-stage focus of general public concern, is a recent event attributable to three causes. The first is the writings of a few inspired and passionate scientists with the gift of riveting the attention of the literate layman. For example, Rachel Carson's *Silent Spring* (1962) was enormously influential and was the first introduction of many people to environmental issues. At the same time, the mass media began to appreciate the significance of environmental issues to people and an increasing output of stories in newspapers, on radio and notably on television has been the result. For example, in 1970, the *New York Times* published 150 general articles on environmental issues, 650 on air pollution and 900 on water pollution (Holdgate, 1990).

The second cause is the sequence of environmental disasters which has occurred in the past thirty years or so and which includes the sinking of the *Torrey Canyon* and *Exxon Valdez*, Bhopal, Three Mile Island, Chernobyl, the desertification of sub-Saharan Africa, the repetitive inundation of Bangladesh and the burning of the Amazonian rain-forest; all have been brought vividly to the attention of the world by means of television.

The third reason why the general public is now aware of environmental issues is the effective and often unorthodox activities of committed pressure groups such as Greenpeace and Friends of the Earth. Both seek the active support of concerned citizens in policies which are science based and often tactically implemented by acts of considerable courage which inevitably attract media attention.

In view of this comprehensive publicity, it is fair to say that no one who has access to the mass media can be unaware that we are faced with terrible threats to the global environment and hence to ourselves. Changes in climatic and hence vegetational patterns are already under way. Where they will lead is a matter of intense debate, but it is clear that plants will have to adapt to survive and this principle applies equally to our cultivated plants.

In the 1960s, the International Council of Scientific Unions founded the International Biological Programme (IBP). Its aim was to understand through the synthesis of existing knowledge and through research, the basic processes which support life on the planet, as a means to solving problems of human welfare. One of the sub-programmes of the IBP was concerned with plant genetic resources and one of its activities, jointly with the Food and Agriculture Organisation of the United Nations (FAO), was to hold a technical conference in 1967. Its outcome was the landmark publication *Genetic Resources in Plants – Their Exploration and Conservation* (Frankel and Bennett, 1970), which assembled current knowledge and synthesised ideas on the subject. It helped focus the attention of the scientific community but sadly had little impact in the wider world; a good illustration of the difficulty of communication between scientific specialists and the general public.

The Stockholm Conference on The Human Environment in 1972 was a milestone in stimulating the awareness of governments and policy-formers of the scale and consequences of environmental degradation.

In 1980 a World Conservation Strategy was launched by the International Union for the Conservation of Nature (IUCN) and the United Nations Environment Programme (UNEP). The idea that all ecological systems were interdependent, began to receive general recognition in the concept that there is but a single biosphere for the planet. This global biosphere has many levels of organisation and Man-induced changes in one system or one place, can have consequences for other systems in other places. For example, the recent change in the annual movement pattern of El Nino, the cold Pacific current, has not only disturbed the anchovy fisheries of Chile and Peru, and the economies of these two countries but, it is now becoming clear, is influencing the climate of other parts of the world.

In 1983, the UN General Assembly set up the World Commission on Environment and Development to address the joint issues of environmental conservation and sustainable development. Its comprehensive report – *Our Common Future* (1987) – breaks new ground in setting out in detail the objective of integrating the alleviation of world poverty with world environmental protection. It is not too much to say that it is now widely recognised that the reconciliation of these aims which traditionally have been mutually exclusive, is the greatest challenge facing mankind.

The prospect is daunting but there are some encouraging signs. For the first time, people are beginning to see themselves as a species with a unique and central role and responsibility in the biosphere and to realise that their selfish manipulation and exploitation of parts of it have profound consequences for the whole and for their own posterity. Happily, an increasing proportion of the general public, and, through it, of governments, is beginning to accept this responsibility and with it the idea that the natural world is not an enemy to be subdued nor a resource provided for our indiscriminate exploitation but rather a partner whose survival is essential to our own.

The problem in an issue of such complexity, is how to turn concern into effective action. One of the obstacles to progress is the gulf which exists between the level at which principles are formulated and internationally accepted, such as those defined in *Our Common Future*, and the level at which they are translated into actions by the industrialists, farmers and foresters of our societies. In the final analysis, changes are dependent on the actions of individuals who are struggling to maintain or improve their standard of living from within established economic and social structures. Few have the moral and financial independence to step alone outside the system, and initiatives must come from elsewhere.

Action plans, if they are to succeed in changing the direction and momentum of established systems, need to be widely accepted, securely founded on reason rather than emotion, and based on adequate scientific and technological information. All too often these elements are lacking. For example, few countries have reliable weather data for as recently as one hundred years ago, and even fewer have completed inventories of plant species or detailed systematic surveys of the composition of the ecosystems within their borders. Consequently, assessment of the significance of present trends and predictions of their future consequences are fraught with uncertainties.

It is clear that the protection of the environment and the conservation of its resources is not a matter for unilateral action by national governments. The solution to environmental problems will require a level of international cooperation hitherto unknown in world affairs. It is important to stress that despite the magnitude of the problems this is not a Doomsday scenario. There are very encouraging indications that nations recognise the need and have the will to organise a collective change of direction towards sustainable development. MacNeill (1990) stresses the central role of the United Nations in all of this and refers to suggestions for a UN Environment Council, equal in authority to the Security Council and known perhaps as the Earth Council.

Genetic diversity of crop plants is at risk

In this book we are concerned with one component of the man-induced changes to the biosphere, the destruction of the genetic diversity of our crop plants. The pressures of agricultural and industrial development in the past hundred years have been so powerful and the consequent changes so rapid, that the only effective response to crisis has been to collect and conserve – usually as seeds in a cold store – representative samples of crop diversity. This collecting, which began on a small scale and in an uncoordinated way in the 1920s and 1930s, long before the conservation of the wider environment became a matter of international concern, developed after 1974 into an urgent large-scale operation for rescuing the diversity of all major crops.

We intend to show why it is important to conserve the genetic resources of our crops, that this work is grounded on science, that it has great economic implications, and that it is essential to food security in a changing world and not least in relation to sustainable development. We will show what has been achieved already, what remains as a challenge for the future and how this challenge may best be met. We attempt to sum up the present status of all aspects of crop genetic resources conservation from the standpoint of its science base and operational effectiveness.

The experience gained in the last fifteen years of urgent rescue collecting has provided many lessons in cost-effectiveness in the planning and execution of genetic resources conservation and in forming cooperative links between people of disparate interests. Three general principles emerge:

- The need for planning and for operational decisions to be based on prior analysis of existing data rather than on inspired intuition.
- The need for the focus of activity to be the crop species or species group rather than the country or region.
- The essential need for active international collaboration.

Finally, we confidently predict that the knowledge gained from collecting and conserving crops and their wild and weedy relatives will often be relevant to the conservation of other plant species.

2

The genetic resources of crop plants

Résumé

Crop plants evolved in parallel with human societies and most of our major crops have spread with man over large areas of the globe. In so doing they have differentiated into a very large number of landraces which are the agricultural equivalent of ecotypes in wild species. This phase of spread and differentiation lasted about 10 000 years and culminated in the maximum expression of genetic diversity. During the past hundred years or so, the process has been arrested and reversed by the intervention of plant breeders who discovered how to produce more uniform and higher yielding varieties. These varieties, which were rapidly taken up by farmers, replaced the landraces in cultivation and so led to the erosion and loss of the natural diversity of our crop plants.

Breeders who had already begun to collect plant material from other countries as sources of new genes, became aware that the success of their work was creating a serious loss of diversity and began to collect for conservation for use by themselves and by posterity.

As the rate of loss accelerated with the increasing effectiveness of breeding it became clear that individual efforts were unable to cope with the scale of the problem and international collaborative action took their place, much of it organised by the International Board for Plant Genetic Resources (IBPGR).

What are the genetic resources of crop plants?

The genetic resources of a crop consist of the total pool of genetic variation which exists in the crop species. This includes the genes of the cultivated, weedy and wild related species which are sufficiently close to the cultivated forms for gene exchange to be possible. Gene exchange may occur by natural or manual crossing, or more recently by manipulative techniques which permit barriers to sexual crossing to be circumvented (for example, by the artificial culture of embryos which would otherwise abort), and in this way the boundaries of the gene-pool can be extended. The limits to effective hybridisation enclose the gene-pool of the crop species group, and the total genetic variation which it contains represents the genetic resources of the crop.

The conservation of crop genetic resources is motivated by practical considerations. It is not conservation for its own sake but for use in crop development and food production, for the present and in the future. Genetic diversity is essential if plant breeders are to adapt crops to meet the changing demands of growers and consumers and to cope with changes in the environment. The production of varieties with characters or combinations of characters different from those in use in the farming systems of today, will depend on the identification and incorporation of genes or gene-systems with the capacity to develop the required characters. These genes must be found in the gene-pool – since with our present state of knowledge they cannot be created – and, having been found, must be incorporated into suitable genetic backgrounds, either by conventional breeding or by the newer techniques of genetic engineering.

Crop plants and human societies evolved together

The origins of agriculture can be traced back, from archaeological evidence, to the Neolithic period, some 10 000 years ago. Prior to this time, and in some areas in the present day, man subsisted as a hunter of wild animals and gatherer of wild roots, fruits and seeds.

The development of the practice of agriculture was a complex process in which the first critical step was possibly the establishment of settlements in order that crops might be sown, guarded and harvested. As soon as the purposive culture of plants began, so too must have begun the process of their domestication. As plants became more adapted to man's needs, so farming became more productive, settlements thrived, populations increased and agriculture spread from its centres of origin into new regions and habitats. Conversely, domestication made the cultivated forms less fit for survival in the wild, and they became dependent on agriculture. Human societies, crop plants and domestic animals evolved together, and were interdependent in both their origins and development. Agriculture enabled humans to establish permanent dwellings, to plan food production, to conserve supplies for non-productive seasons, to develop organisation in society through division of labour, and, with material needs more secure, to develop their unique faculty for intellectual pursuits. Agriculture has been equated with the discovery of fire and the invention of the wheel in its significance to human development. However, it has been both more complex and more far-reaching than either of the other discoveries.

The culture of crops undoubtedly began independently in several places at different times. The number and location of these cradles of agriculture has been the subject of much debate which we do not propose to join, but for a concise review see Zeven and de Wet (1982). Harlan (1971) proposed three primary centres: east Asia, Asia Minor and Meso-America, whereas Hawkes (1983) proposed four: northern China, Asia Minor, southern Mexico and central to southern Peru.

The early cultivators naturally used whatever plants seemed most suitable to them from those available from the local flora. As agriculture spread from the primary centres the primary domesticates would have spread with it, and new wild species were domesticated along the way. The probable areas of origin of some of the major crops are given in Table 2.1. It is interesting that, of the enormous diversity of species available at each centre, relatively few were domesticated and that world-wide, of an estimated 200 000 species of flowering plants, only about 300 have been used in agriculture. Of these, twenty to thirty species provide most of man's food (Harlan, 1976). On the other hand, the villagers of Indonesia are said to cultivate as many as 3000 species of fruits, vegetables, spices and medicinal plants in their gardens (Hardon, pers. comm.). While the scale of production is tiny, the diversity suggests great potential for wider utilisation. Some crops, such as wheat and maize, have achieved a world-wide distribution; others have remained highly localised, such as quinoa in the Peruvian Andes. The reasons for these differences are not clear but possibly lie in innate differences in productivity and in differences in the adaptability of the species to new habitats. Crops differ in the degree of their domestication; contrasting examples are sugar beet – a modern, virtually man-made crop, and many tropical tree fruits, which differ little from wild species.

As crops spread from their primary centres of origin into new habitats differing from the centre and from each other, they were subject to differing selection pressures, resulting in the development of different agro-ecotypes, or landraces. In this way, the spread of agriculture was accompanied by the evolution of diversity. In some cases, where a cultivated species reached the limits of its adaptive range, another species which had been previously

Table 2.1. *Probable areas of origin of major crop plants (From Harlan, 1976.)*

Andean South America	*Africa*	*South Pacific*
Potato	African rice	Sugar cane
Peanut	Sorghum	Coconut
Lima bean	Pearl millet	Breadfruit
Cotton	Finger millet	*South-east Asia*
Squashes	Yam	Asian rice
North-east South America	Cow pea	Banana
Yam	Coffee	Citrus
Pineapple	*Asia Minor*	Yam
Cotton	Wheat	Mango
Sweet potato	Barley	Sugar cane
Cassava	Onion	Taro
Meso-America	Pea	Tea
Maize	Lentil	*China*
Tomato	Chick pea	Soybean
Cotton	Fig	Cabbage
Avocado	Date	Onion
Papaya	Flax	Peach
Cocoa	Pear	
Cassava	Olive	*Central Asia*
Sweet potato	Grape	Common millet
Common bean	Almond	Buckwheat
North America	Apricot	Alfalfa
Sunflower	*Europe*	Hemp
Cotton	Oats	Broad bean
	Rye	*India*
	Sugar beet	Pigeon pea
	Cabbage	Eggplant
		Cotton
		Sesame

associated with it as a weed, became the dominant component of the crop–weed mixture and evolved as a new, substitute crop. This seems to have happened to oats in wheat in north-western Europe, where the oat component thrived better under the milder and moister conditions than the wheat which it was infesting. In central Europe, *Fagopyrum tataricum* grows as a weed in buckwheat *F. esculentum* crops, but at high elevations in central Asia, *F. tataricum* replaces *F. esculentum* as the more productive species.

The development of diversity which accompanied the spread of crops was confined to the cultivated species of a crop group. Wild forms remained in their natural habitats except in the case of those species which took part in the migration as weedy components of the crop. The adaptation of wild species was to the ecology of their natural habitats, whereas in the case of cultivated and weedy forms, where man was responsible for extending their range, the adaptation was agro-ecological, that is, adapted to agricultural habitats and to farming practices.

The domestication of crops refers to the development of new forms of the plant which are more fitted for cultivation and use. The motive force for domestication is selection pressure, exerted by the cultivator and the cultivation process, on natural variants. Examples of adaptive changes brought about by domestication include:

Adaptation of seed for easier sowing, often through the loss of hairs and spines on the seed coat.

Greater uniformity in crop growth and ripening resulting from uniform germination following loss of seed dormancy mechanisms.

Improved harvesting efficiency from the elimination of spontaneous shattering of the inflorescence or fruit with consequent seed dispersal.

Improved edibility from the elimination or reduction of toxic principles in that part of the plant which is used for food.

Domestication changes seem to have been greater in crops grown for their seeds, the cereals and pulses, than in roots and vegetables such as potatoes, yams and cassava, and has been least in fruits, particularly in tropical tree fruits. In all cases the changes can be seen to reflect the needs of the cultivator and consumer; the greater the degree of domestication – the greater the change from the wild types – the greater the dependence of the domesticate for survival on man and of man on the domesticate for food or other use.

Crops evolved from wild species

Most crops are of great antiquity

The areas of origin of ancestral wild species, and their evolutionary pathways to the cultivated species have been the subject of intensive study, particularly for the crops of major economic importance. Evidence comes from four principle sources:

1. From archaeology. Plant remains, which can be dated by means of radio-carbon methods, have been found at sites of early human habitation. This type of evidence is most likely to come from sites in dry climates and for plant parts which are relatively resistant to decay, particularly dry seeds and those carbonised by proximity to fire, and from artifacts such as the woven fibres of cotton and flax.

2. From comparative studies of the form and structure of durable plant parts: seeds, dry fruits and dispersal units.

3. From chromosome studies. Chromosomes contain discrete parcels of DNA which are the genes. They exist in the nucleus of every cell. Chromosomes can be stained and counted under a microscope and they have characteristic sizes and shapes. Each species has a characteristic number of chromosomes which embody its basic set of genetic information, the genome. Most plant species have two genomes per cell and are said to be diploid, but polyploid species with four, six and eight genomes occur. In the process of egg and pollen grain formation which precedes sexual reproduction, chromosomes pair together according to their genetic similarity. The affinity between the chromosomes of different species can be assessed by the degree to which they pair in hybrids. Complete pairing indicates a close relationship, and absence of pairing implies genetic distance. Genomic similarities and differences are indicated by alphabetical symbols. Thus the two tetraploid wheat species *Triticum dicoccoides* and *T. timopheevii* have genomic constitutions of AABB and AAGG, respectively, indicating that they have the A genome, and hence one diploid ancestor in common, and that the other genome and ancestor is different in each case.

4. From historical records, which, though lacking for the critical period of 10 000 to 2000 BP, do provide some information for the period 2000 to about 400 years ago. Diaries of navigators, explorers and colonists provide information on crops discovered, and those actively transported between the Old and New Worlds, in more recent times.

Fascinating histories of crop origins are emerging. Huge uncertainties remain, particularly for the dates and locations of primary domestications and rates of spread, and the degree of understanding varies greatly from one crop to another. However, for a number of major crops, it is possible to sketch reasonably clear pictures of their evolution and spread.

Banana

There are twenty-nine to thirty-nine species in the genus, according to different authorities. Only two species are important in the evolution of the edible banana.

The key species is wild *Musa acuminata*, with a genomic formula AA. *M. acuminata* produces inedible fruits with many seeds, and little flesh. In the domestication of the banana, the two critical steps were the occurrence of female (though not male) sterility, and the development of partheno-carpy (Simmonds, 1976). Female sterility removed the seeds and partheno-carpy caused the development of the fleshy fruit tissue independent of the normal stimulus provided by developing seeds. These two events probably occurred as natural variants in *M. acuminata*, possibly in the Malayasian peninsula. The first edible types were no doubt subject to strong positive selection, were mulitiplied vegetatively to form clones and were dispersed by migrating people throughout south-east Asia. Thus, wild bananas were transformed from a tropical forest constituent to a productive crop.

The spread of banana cultivation would have brought the *acuminata* clones into the distribution range of the second critical species, *M. balbisiana* (genome BB), which occurs from India to the Philippines and New Guinea, but not in Malaysia. Natural hybridisations followed, presumably with *M. acuminata* as the male parent, to give hybrids with genomic constitution AB and a series of polyploid forms with various genomic complements, such as triploids AAA, AAB and ABB and tetraploids AABB, AAAB and ABBB. *M. acuminata* originated in the humid tropics, and *M. balbisiana* is adapted to areas with a monsoonal climate, which includes a pronounced dry season. The various combinations of the two different genomes therefore provided a range of clones with the capability of extending the range of banana cultivation beyond that possible with the *M. acuminata* genome alone. In addition, *M. balbisiana* provided valuable genetic contributions to disease resistance and to fruit characters such as dry matter, starch, acid and Vitamin C contents as well as to texture and flavour. Triploids are more vigorous and have larger fruits than diploids, and constitute the majority of the 500 or so clones which are known today.

According to Simmonds (1976), no dates can be assigned to the early evolution of the bananas but it is likely to have happened thousands rather than hundreds of years ago. Spread of bananas eastwards across the Pacific accompanied the Polynesian migrations. Westward migrations were probably directly from Indonesia to Madagascar, thence up the Zambesi valley and via the great lakes and Congo to West Africa. The vegetative nature of the propagating material and its sensitivity to dessication and decay must have exerted a powerful influence on both the route and the rate of migration across seas, much if not all of which must have been in open canoes and small boats. In traversing central Africa from east to west, the crop would have encountered suitable ecological conditions along the route (Purseglove, 1985b). Portuguese explorers record finding bananas in west Africa in the 15th century and took them to the Canary Islands. The first record of an introduction to the Caribbean was in 1516, from the

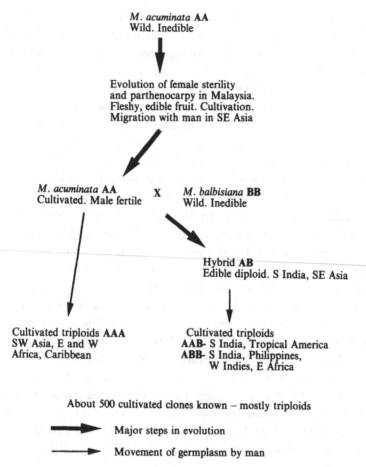

Fig. 2.1 The origins and distribution of cultivated bananas (*Musa* sp.). (Adapted from Simmonds, 1976, and Purseglove, 1985b.)

Canaries. Early introductions into east Africa are thought to have been AAA clones, and recent production in Uganda is based on clones of this type, as is production in the various other countries along this route westward to the Caribbean. The origin and distribution of the different genomic combinations is illustrated in Fig. 2.1. It should be stressed that this distribution is primarily dependent on the movement of clonal material by man and that what we see now is the final outcome of these migrations after selection among the migrant clones for their adaptation to new environments and for suitability for human needs.

It is important to emphasise the very narrow genetic base of the banana crop. Seedlessness (female sterility) means the loss of sexual reproduction and the only remaining possibility for genetic variation in the cultivated varieties lies in the accumulation of naturally occurring mutant genes. The extreme genetic uniformity of the crop makes it subject to major epidemics of diseases such as banana wilt, (Panama disease), leaf spot (sigatoka) and bunchy top, an insect-transmitted virus disease, and to many insect pests. This has prompted Simmonds (1976) to describe the banana as 'the best example in the history of agriculture of the pathological perils of monoclone culture'.

Cotton

Thirty-four species have been described in the genus *Gossypium*, thirty diploids and four tetraploids. Of these, four species, two diploids and two tetraploids, have been involved in the evolution of the lint-producing cottons.

G. herbaceum (AA) is widely distributed as a wild perennial plant in the savannah vegetation of southern Africa. It is thought to have been cultivated first in Syria and Arabia before being carried along trade routes into the Indian subcontinent where *G. arboreum* arose from it under cultivation. No wild form of *arboreum* is known. In its turn, *arboreum* was dispersed along trade routes and became the dominant type of cotton in Asia and Africa. Fragments of cotton cloth from 5000 years ago have been found in the Indus valley.

The two significant tetraploid species, *G. barbadense* (AADD) and *G. hirsutum* (AADD), are both New World in origin. They are thought to have arisen from a hybridisation between plants of two diploid species carrying the A and D genomes, respectively. The D genome donor has been identified as *G. raimondii*, a species indigenous in Peru. The difficulty is to understand how the A genome of *G. herbaceum* travelled more than 5000 years ago from Africa to South America, since no A genome diploids are known in the New World. Long-distance seed dispersal is a possibility since cotton seeds can retain their viability after long immersion in sea water, and the transport of seeds by ocean currents from Africa to South America may have occurred.

Whatever the precise origin of the A genome in the New World, undoubtedly a hybridisation between a diploid *G. raimondii* type and an A genome carrier, followed by a chromosome doubling event, gave rise to the first wild tetraploid (AADD), whose descendants, in migrating from their centre of origin, differentiated into the two perennial New World tetraploids, *G. barbadense* in north-west and north-east South America and

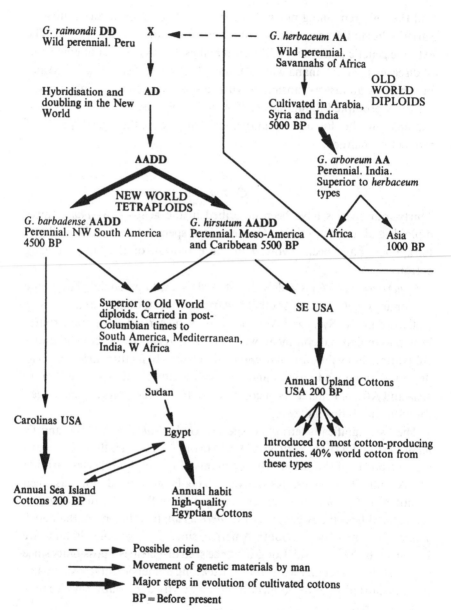

Fig. 2.2 The evolution of cotton (*Gossypium* spp.). (Based on Phillips, 1976, and Purseglove, 1982c.)

G. hirsutum in Meso-America and the Caribbean. Evidence of the domestication of *G. hirsutum* in Mexico goes back some 5500 years, and of *G. barbadense* in Peru, about 4500 years. Cotton was widely grown in Central and South America in pre-Columbian times.

Perennial forms of *G. barbadense* and *G. hirsutum* were carried by the Spanish to the Mediterranean and by the Portuguese to Africa and India, where, because of their superiority, they often replaced the indigenous diploids. The final stage in the evolution of modern cotton varieties was the independent origin of three different annual types of New World cottons. Upland cottons (derived from *G. hirsutum*) which came into cultivation in the mid 18th century in the south-eastern USA; Sea Island cotton (from *G. barbadense*) in the coastal areas of the Carolinas and Georgia, USA, in the late 18th century; and Egyptian cotton, derived partly from Sea Island germplasm introduced from the USA in the late 18th century. In turn, some of the products of Egyptian breeding and selection were introduced into the south-west USA, and according to Phillips (1976) contributed as parents to the breeding of improved varieties in that area (Fig. 2.2).

This free movement of species and varieties between countries and continents for direct use as crops and for use as parents in breeding programmes, is a frequent event in the evolutionary history of many crop plants and in this century has become commonplace as the number and scope of breeding programmes has increased. This has resulted in gene exchange between previously distinct types and a blurring of differences. It also means that, as improved varieties come to be widely grown, there is a narrowing of the genetic diversity of the cultivated crop. Today, about 95 per cent of world production is from annual *G. hirsutum* types, and the remainder is from *G. barbadense*.

Wheat

The evolution of bread wheat has involved four species of wild annual grasses, all indigenous in Asia Minor (Feldman, 1976). The tracing of the evolutionary pathway has required intensive study of genomic affinities of a large number of species in the genus *Triticum* and the related genera *Aegilops* and *Haynaldia*, all members of the tribe *Triticeae*. The essential steps and probable critical species in the evolutionary sequence are shown diagrammatically in Fig. 2.3. The steps include natural hybridisations in the wild, chromosome doubling events, from diploid to tetraploid and from triploid to hexaploid, with domestication occurring at each chromosome

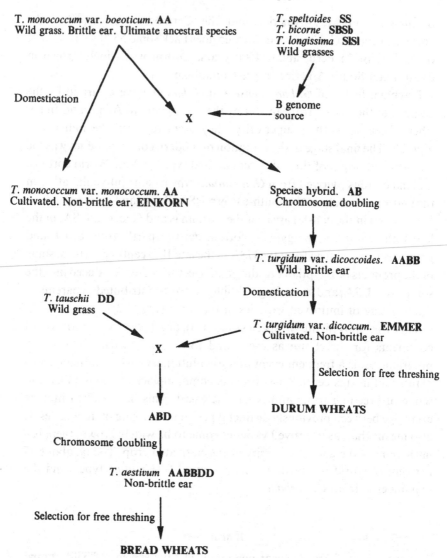

T. *monococcum* var. *boeoticum.* **AA**
Wild grass. Brittle ear. Ultimate ancestral species

T. *speltoides* **SS**
T. *bicorne* **SBSb**
T. *longissima* **SlSl**
Wild grasses

Domestication

B genome
source

X

T. *monococcum* var. *monococcum.* **AA**
Cultivated. Non-brittle ear. **EINKORN**

Species hybrid. **AB**
Chromosome doubling

T. *turgidum* var. *dicoccoides.* **AABB**
Wild. Brittle ear

T. *tauschii* **DD**
Wild grass

Domestication

T. *turgidum* var. *dicoccum.* **EMMER**
Cultivated. Non-brittle ear

X

Selection for free threshing

ABD

DURUM WHEATS

Chromosome doubling

T. *aestivum* **AABBDD**
Non-brittle ear

Selection for free threshing

BREAD WHEATS

Fig. 2.3 Evolutionary pathways of cultivated wheats (*Triticum* spp.). (Modified from Feldman, 1976.)

level to give cultivated diploids and tetraploids as well as the cultivated hexaploid bread wheat.

The sequence began with *T. monococcum* var. *boeoticum* (AA), a large-grained wild grass with brittle ears, that is, with ears which spontaneously break up when mature, as part of the seed dispersal process. The earliest finds of carbonised grains of *T. monococcum* var. *boeoticum* were in association with ancient settlements in the mountainous Fertile Crescent around the Tigris and Euphrates river systems, and are about 10 000 years old. These grains may have been collected from the wild natural stands of the grass, but during the next millenium, non-brittle grains appear among the finds and subsequently replace the brittle types. This signals the substitution of deliberate cultivation for gathering from the wild, because the non-brittle character is clearly disadvantageous to survival in the wild while retention of the intact ear on the stalk is an obvious convenience to the harvester. It seems likely that there was extreme selection pressure in favour of the non-brittle type, *T. monococcum* var. *monococcum* or Einkorn, which in succeeding centuries spread widely through the Balkans and Europe. Today it occurs as a localised relic crop in mountainous areas of Turkey and Yugoslavia.

The critical wild tetraploid species is *T. turgidum* var. *dicoccoides* (AABB) with a brittle ear. It gave rise to the non-brittle form *T. turgidum* var. *dicoccum*, or Emmer, perhaps contemporaneously with the origin of Einkorn. Emmer spread widely in Asia Minor and later to Europe, the Mediterranean, India and central Asia. The origin of its A genome is confidently attributed to *T. monococcum* var. *boeoticum* but the source of the B genome has so far defied unequivocal identification. Studies of chromosomes and of the morphology and geographical distribution ranges of the possible progenitor species point to *Triticum speltoides* (SS) as the most likely donor of a genome which subsequently became modified to BB, either before the hybridisation with *T. monococcum* var. *boeoticum* or perhaps, more likely, subsequent to it and to the differentiation of *T. dicoccoides*.

The final step in the synthesis of the genetic complement of bread wheat, *T. aestivum* (AABBDD), occurred when a cultivated *T. turgidum* var. *dicoccum* (AABB) (crossed with *T. tauschii* (DD). The cultivation of *T. turgidum* var. *dicoccum* has commonly occurred within the natural distribution range of *T. tauschii* and the latter would doubtless have been present either as a weed within the crop or around the margins of cultivation. The absence of any wild hexaploid progenitors of *T. aestivum* supports this view of an origin in cultivation. All known forms have a non-brittle ear. It follows that *aestivum* evolved after the cultivation of Einkorn and Emmer, but even so the earliest finds date back to about 9000 years ago. The final major step in the domestication of both Emmer and *aestivum* was the selection of genetic mutants which confer the 'free-threshing' characteristic whereby the grain can be easily separated from its enveloping chaff and the texture of the flour thereby improved. This complex sequence of species hybridisations, genomic doublings and domestications has led to the development of our present-day tetraploid durum or macaroni wheats and the hexaploid bread wheats.

The degree of domestication is such that neither tetraploids nor hexaploids can survive without the intervention of man in their cultivation. The dependence of man on the crop can be judged from its occurrence as a major staple from Scandinavia to Argentina. In the tropics it is confined to high altitudes. The principal producing regions are the southern former USSR, the central plains of North America, the Mediterranean basin, China, India, Argentina, southern Australia and Europe. It is a major crop of world trade, which in 1983 was grown on about 230 million hectares (ha) with a world production of about 500 million tonnes (t). All this from a crop which originated from wild grasses in a relatively small area of Asia Minor. The range of variability released from the reassortment of its genetic constitution has given rise to an estimated 17 000 different varieties in addition to uncounted landraces.

A timetable of the origin and spread of the three cultivated wheat groups is presented in Table 2.2.

Rice

Cultivated rices fall into two groups similar in appearance but distinguished by their different areas of origin and by the sterility of hybrids between them, indicating significant genetic separation. They are, *Oryza sativa*, Asian rice, and *Oryza glaberrima*, African rice. Both are presumed to have descended from a common ancestor, perhaps *O. perennis*, a wild perennial species widely distributed throughout the tropics. The evolution of the two groups proceeded independently with similar steps in each case; first, the evolution from the presumptive common ancestor, of two wild perennials, one Asian the other African. Each in turn gave rise to a wild annual species from which the cultivated annuals arose (see Fig. 2.4). Chromosome studies show that the species in the two pathways share the same genome AA, which appears to have undergone some modifications in the *O. glaberrima* pathway. These differences are associated with, though not necessarily responsible for, the genetic isolation between the two groups. Uncertainties exist about the precise identity of ancestral species, partly because of disagreements on the naming and identification of species and partly because of the absence of major genomic differentiation, which provided strong evidence of origin in other species groups such as wheat and cotton. All cultivated rices are diploids (Chang, 1976, and Purseglove, 1985a).

Table 2.2. *Times of origins and geographical dispersions of cultivated wheats*

Years before present	Einkorn	Emmer	Bread wheat
10 000	Fertile Crescent		
9000	Balkans	Asia Minor, SW Asia	Syria
8000		Mesopotamia	Iran, Iraq, Anatolia, Crete, Mesopotamia
7000	Danube and Rhine valleys	Egypt	Egypt, Mediterranean
6000	C and W Europe, Asia Minor	Mediterranean Basin	C and W Europe
5000		Europe, Ethiopia, C Asia	Indus valley, China, C Asia
4000		India	
		↓	
		Replacement by free-threshing Durums in most areas not invaded by bread wheats	
		↓	
300–100			Mexico, Australia, USA, Argentina, Canada
Today	Relic cultivations in Yugoslavia and Turkey	Relic cultivations in Iran, Ethiopia, E Turkey and Balkans	

Notes:
Dispersion by man of Emmer, Durum and bread wheats led to differentiation of tens of thousands of locally adapted landraces, now largely replaced in cultivation by modern varieties.
Earliest records obtained from carbonised or dried grains from prehistoric sites.

O. glaberrima occurs in sub-Saharan Africa, in flood-plain habitats where it is sown before the floods arrive and harvested after they have receded. Upland or 'dryland' varieties are also grown. Although it is hardier than *O. sativa*, it nevertheless has been replaced by *O. sativa* wherever conditions permit the latter's cultivation. The greatest diversity of *O. glaberrima* is in West Africa.

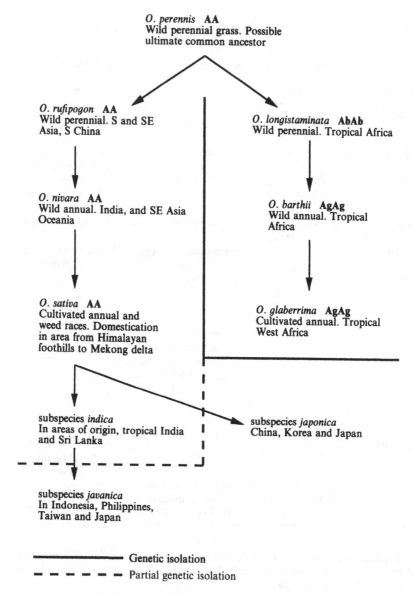

Fig. 2.4 The independent evolutionary pathways of Asian (*Oryza sativa*) and African (*O. glaberrima*) rices. (Modified from Chang, 1976.)

The starting point for the differentiation of the Asian rices is the perennial wild grass *O. rufipogon*, widely distributed in south and south-east Asia, south China, and South America, from which the annual wild species *O. nivara* arose. *O. nivara* has a wide distribution range in India, south-east Asia and Oceania. Domestication of *nivara* involved selection for non-brittle ears, larger plants and grain, and shorter dormancy, and it probably occurred, perhaps in several independent cases, within the large area of north-east India, northern Bangladesh, Burma, Thailand, Laos, Vietnam and south-east China. From this diffuse area of origin, cultivated rice was rapidly and widely dispersed into new regions and habitats by farmers who doubtless could see the immense benefits to be gained from the addition of a cereal crop to a farming culture which up to that time had lacked one.

Man carried the early domesticates into different climatic zones, into areas of different daylengths and on to different soil types, and selected thousands of locally adapted landraces. These fall into three sub-species: ssp. *indica*, in the area of origin and throughout India and Sri Lanka, ssp. *japonica*, the temperate rices of China, Japan and Korea, and ssp. *javanica* in Indonesia, the Philippines, Taiwan and Japan.

This expansion of rice cultivation by man, southwards and eastwards from its area of origin, is thought to have been facilitated by frequent natural crossing between the cultivated forms and their associated wild and weedy species which provided additional sources of genetic variability for selection. There is little information on the early history of Asian rice culture to help us understand the route and timing of its dispersion from its presumed centre of origin in the sub-Himalayan arc of south Asia and, according to Chang (1976), much of what is known is controversial.

In what follows, the dates given are often based on archaeological finds, for example, on impressions of rice chaff on pottery, and are isolated data which indicate only the earliest known cultivation in a particular area and not necessarily the time when cultivation first began.

In India the earliest evidence goes back 4500 years and in south China records indicate that rice was cultivated 5000 years ago. From south China it was taken to the Philippines by 3500 BP and to Japan about 2000 BP. Another dispersion route was down the Malay archipelago and thence to Indonesia by 3500 BP. Whatever the precise routes and times, rice clearly spread widely and rapidly southwards and eastwards through tropical and sub-tropical Asia together with its associated weed races.

It also moved westwards, reaching Asia Minor and the Nile valley by 2000 BP, and the western Mediterranean soon afterwards. The Portuguese or Spanish voyagers took it to South and Central America by 400 BP. It reached the USA about 300 BP and finally Australia some 200 years ago (Purseglove, 1985a).

Dispersion of crops depends on adaptation to new habitats

Adaptation has a genetic base

The idea that the adaptation of plant populations to their habitats has a genetic and therefore a heritable base, is well established in the science of genecology. Experimental work has shown how natural selection has operated on genes and gene combinations to favour those which are most fitted to the local environment, and since these combinations contribute disproportionately to the genetic constitution of the following generations, a type adapted to the local ecology – an ecotype – will develop. As the natural dispersal mechanism of the species takes it into different habitats, the process is repeated, and so long as the naturally available genetic diversity is sufficient to permit the generation of new adaptive combinations, the dispersion will continue and the species will come to consist of a series of ecotypes. For a discussion and experimental proof of ecotypic differentiation in action see Bradshaw (1975).

Differences between ecotypes are often subtle and may be difficult to detect. Furthermore, each ecotype is not genetically uniform. Individuals in wild populations are typically cross-pollinating and consequently retain genetic diversity and the population remains heterogeneous. This residual genetic diversity provides the opportunity for further response within the ecotype to subsequent change in the environment. Few environments are stable over time; annual fluctuations in rainfall or in winter temperatures are common examples of instability to which ecotypes are constantly fine-tuning. If we compare ecotypes from widely different parts of the species distribution range, they may be qualitatively different; some genes present in one may be lacking in the other, but usually the differences will be quantitative, with differing frequencies of particular genes in the populations.

Landraces of crops are equivalent to ecotypes of wild species

The same principles apply to adaptive change in primitive cultivated species. Here, however, we have additional selection pressures to consider. Adaptation to different cultivation practices, for example different sowing and harvesting dates, becomes a major selection pressure, whether applied consciously or unconsciously by the farmer. When the movement of a crop by man is sudden and into different latitudes, rapid responses can be required.

In many crops the initiation of flowering is triggered by daylength. This is true also of tuber formation in the potato. Potatoes were first brought to Europe by returning Portuguese and Spanish explorers from the north coast of South America, perhaps as stores for the voyage. The species found its way into the botanical gardens of Europe, where it existed as an exotic specimen for a hundred years or more before its potential as a European crop was appreciated (Burton, 1966). The reason for this apparent lack of enterprise, with European farmers failing to exploit the potato's potential, lies primarily in the crop's origin in equatorial latitudes where daylengths are short. The introduced South American material would not have produced tubers until the European daylengths were similar to those in the centre of origin, that is until the European autumn. Yields, therefore, were undoubtedly low and tubers small. On the other hand, distribution of seed among European botanists provided the opportunity for the culture of genetically diverse progeny and the selection of plants capable of starting tuber formation under the long days of summer. Adaptation of the European potato took place first of all in botanic gardens and subsequently under farming conditions to produce a range of varieties adapted to northern temperate conditions and to local farming and consumer requirements.

The landrace phase of crop development culminates in the maximum expression of within-crop diversity

The examples of wheat, rice, cotton and banana discussed above, with their distribution ranges extending across continents, oceans and hemispheres, illustrate the great diversity of habitats to which migrating crops have successfully adapted. It is worth reiterating that this evolution of crop diversity was dependent on there being genetically diverse forms on which selection could act. It was such an astonishingly successful process precisely because there were the genetic resources available for the emergence of adapted ecotypes

For many temperate annual crops this process of diversification seems to have reached its peak at the middle to end of the 19th century. For tropical crops it was probably later, but for all major crops diversification was at a maximum by the end of the landrace phase of crop development. This process is now in reverse. The diversity of landraces which supported agriculture for the past 9000 years is being rapidly eroded and, for some temperate crops, is now nearing completion. This has happened through the substitution of new, genetically uniform cultivars which are grown in environments which also have become more uniform through the application of increasingly sophisticated agronomic practices, including improved tillage, irrigation, artificial fertilisers and the chemical control of pests and diseases.

The rapid rate of destruction of crop variability is in sad contrast to the rate of its creation – about 100 years compared to 5000 to 10 000 years.

Plant breeding has resulted in genetic erosion as well as crop improvement

The first cultivars were simple selections for yield and uniformity

As part of a European movement for the improvement of agriculture, a few outstanding farmers and landowners, with the time and resources to experiment, began, in the early 19th century, to practise selection on the variability within the landraces of crops which they were growing. Landraces characteristically consisted of mixtures of plants differing in form, colour, height, vigour and yield. From this diversity they selected single plants which they considered to be superior and multiplied their progeny in isolation. They assessed the merits of these selections against each other and against the landrace from which they were derived. The outcome was the production and marketing of improved varieties which frequently bore the selector's name. Two notable names in this phase of crop development were Vilmorin in France and Shirreff in Scotland; both produced new varieties of wheat which replaced the landraces from which they were derived. Rimpau, in Germany, did comparable work on rye but he selected a number of similar plants to form the nucleus of the new variety. This was the mass-selection method, in contrast to the single-plant selection of the wheat improvers. With the benefit of present-day genetical knowledge, we can see that both methods were appropriate to the different crops. Plants of self-pollinating species, like wheat, have a low level of genetic variability, give rise to progeny like themselves and can be satisfactorily propagated from a single individual. This explains why the early selectors were so successful in producing superior uniform varieties of wheat – and also of inbreeding oats and barley. Rye, like most wild species, including the wild ancestors of the self-pollinating cereals, is a natural outbreeder and plants have a naturally high level of genetic diversity. They need to be propagated from a group of parental plants in order to maintain vigour in the progeny. Since Rimpau's new varieties were often based on twenty or more plants, they were vigorous though perhaps more variable than the new wheats.

The early improvers of wheat, oats and barley found that progress through further rounds of selection rapidly came to a halt. Their success in producing uniform new varieties necessarily reduced or eliminated genetic variation and hence the possibility of finding further variation on which progress depended. Thus, as landraces were supplanted, genetic erosion of the gene-pool began.

Hybridisation generates new variability

In 1761, Koelreuter in Germany published the first report of a plant hybrid produced in a scientific experiment. He crossed two species of *Nicotiana* and raised some twenty hybrids, all of which were sterile. However, he went on to produce an astonishing range of plant hybrids between species in many genera and from this work developed a coherent theory of sexuality in plants and of the significance of hybrids. Acceptance of his ideas was slow, even in the scientific community, and did not influence the agricultural plant improvers until the beginning of the 19th century.

At this time Knight and Herbert in England and Von Gartner in Germany were experimenting with the production of hybrids and were selecting among their progeny in programmes of controlled plant breeding. This work pointed the way forward to selectors like Shirreff and Vilmorin when they found the simple selection in their improved inbred cultivars brought no further progress. They began breeding from carefully selected parents, they selected among the variable progeny, and crop plant improvement entered its most dramatic phase. A theoretical basis for their work was provided by the subsequent discovery and development of Mendel's work on inheritance at the beginning of this century.

Towards the end of the 19th century a great surge occurred in the breeding of crop plants, among scientists concerned with agricultural improvement but also to a remarkable degree among farmers and gardeners. The outcome was a flood of new varieties accompanied by unsubstantiated claims of excellence from their breeders. Confusion reigned about the merits and identities of the varieties. Not infrequently the same variety was sold under different names in different areas. Eventually, order emerged and legislation controlled the authenticity and purity of seeds offered for sale. With the development of statistical theory and its application to field trials, objective data replaced biassed optimism in the assessment of new varieties. Agricultural production came to be based on fewer and fewer high-performing varieties and this trend has been reinforced in some countries in recent years by regulations limiting trade in seed to officially approved varieties.

As genetic resources are destroyed, the need for them increases

In almost all crops there is a general trend towards the production of higher yielding more uniform varieties and their use over wide areas in which they replace the diverse heterogeneous landraces. The rate of this change varies from crop to crop and region to region but the consequences are the same for all crops; a progressive loss of the genetic diversity which had evolved during the past 10 000 years.

Does this loss of old material matter? If the landraces and early cultivars were discarded because they were shown to be inferior to the varieties which replaced them, should we care? The answer is an emphatic, yes!

Modern high-yielding varieties (HYVs) represent the current pinnacle of achievement in adapting crop plants to present needs. However, we should remember that no variety is perfect. Today's best will not be good enough tomorrow. Breeders' objectives are constantly being redefined to meet the changing requirements of farmers and consumers and new challenges from the environment. For example, the use of a restricted range of varieties over large areas approaches genetic monoculture. This has profound effects on the interaction of crop plants and their pathogens, and can lead to epidemics of disease and to calls for new disease-resistant varieties. The major climatic changes foretold as a result of the progressive increases in the levels of atmospheric carbon dioxide and other greenhouse gases will no doubt require further adaptation of crops in existing and new areas of production.

In order to meet their objectives, plant breeders often need to look for genetic variability outside the narrow gene-pool of the current varieties. Since genes cannot yet be constructed at will, the only source to which they can turn is the pool of genetic variability in the varieties and landraces of the past and in related wild species. These classes of plant material constitute a genetic resource which experience shows is of priceless value, and examples of their successful use are given in Chapter 3.

The significance of genetic resources lies in the particular genes or gene combinations which they possess. In general they are used as parents in breeding programmes and therefore they contribute indirectly to crop improvement. The introduction of varieties for direct use in agriculture has had a part to play in the past, usually in the early stages of the development of a crop outside its normal range of distribution. Material for this purpose should be distinguished from genetic resources for conservation because interest in it centres on its immediate suitability and because it is usually discarded if it is found to be of no immediate value.

Modern agriculture in North America, in Australasia and in south Africa was based on varieties of the staple crops of Europe. The success of these introductions was due in part to their genetic heterogeneity which provided the opportunity for adaptive response to their new environments. More recently, the crop-breeding centres of the Consultative Group on International Agricultural Research (CGIAR), have adopted a policy of distributing, for local assessment and selection, large numbers of progeny from their breeding programmes. These materials should be regarded as an intermediate category of germplasm; derived from the use of genetic resources but still unfinished as crop varieties and requiring further selection.

Genetic resources are samples of genetic diversity, usually, but not always, in the form of small samples of seeds. These seeds can provide the parents for breeding programmes aimed at the production of improved crop varieties, but whether they are used immediately for this purpose or not, they are carefully conserved for future use.

N. I. Vavilov, the father of crop genetic resources conservation

No account of crop genetic resources would be complete without due mention of Nicolai Ivanovich Vavilov, 'probably the most distinguished plant breeder, agro-ecologist and applied geneticist of his generation' (Hawkes, 1990). He was appointed in 1920, at the age of thirty-three, as Director of the All-Union Institute of Plant Industry in Leningrad. He was concerned with the study and collection of plants for the improvement of agricultural production in the USSR, through the introduction of new crops and the improvement and the extension of the range of established crops by plant breeding. Initially he was a collector and plant breeder and he displayed astounding energy in the lengthy and arduous collecting missions which he undertook in many parts of the USSR and in south-west Asia, the USA, Canada, the Mediterranean basin, Ethiopia, Eritrea, China, Japan, Korea, Taiwan, Mexico and Central America, the Caribbean and in many countries of South America. He published extensively on the results of his collecting missions and produced detailed monographs on the taxonomy and breeding value of the collected material.

It is said that in the organisation for which he was responsible, he, and his 20 000 colleagues in more than 400 research institutes and breeding stations throughout the USSR, amassed more than 250 000 accessions which they studied, described and carefully stored, irrespective of their immediate utility.

Arising from this essentially practical work, he introduced a number of new theoretical concepts which were to have a profound effect on the thinking of plant breeders and geneticists world-wide. He believed that the area where the greatest diversity of a crop occurred, was its centre of origin and the area of its domestication, and he proposed eight centres of origin for our crop plants. Furthermore, he found from his studies that there were similar patterns of variation in several species groups and in consequence he proposed his Law of Homologous Series. This enabled him to predict the existence of species with particular attributes, which had not yet been found.

His theories have since been criticised and modified by other workers, a normal process in the development of scientific ideas. Their importance lay in the enormous stimulation they provided to the thinking of others and in their demonstration of the need for multidisciplinary studies of crop plants.

For a fascinating general account of this astonishing man and of his work, by one who knew and worked with him, the reader is referred to Hawkes (1990).

Breeders start collecting genetic diversity

Following the lead of Vavilov, breeders in many countries began to assemble collections of material of potential use in their breeding programmes. They did this by exchanging material among themselves and by going, either alone or in joint collecting expeditions, to the areas of diversity of the crops or wild species which interested them.

These collections, started in the 1930s and 1940s, subsequently increased in number and expanded in size as breeders in a number of crops began to feel the constraints of a narrow genetic base. Crops became more susceptible to epidemic diseases and the imperative need for new sources of disease resistance became apparent to all concerned. Collections were regarded as adjuncts to breeding programmes rather than as resources to be conserved for their own sake, and their standard of maintenance, while variable, generally left much to be desired; indeed, not infrequently, material thought to be of little value was discarded. The collecting missions served to highlight the disappearance of natural landraces and primitive varieties, and scientists with a wider vision were, by the late 1960s, becoming increasingly concerned about the accelerating loss of this germplasm, particularly of the major cereals. These had often disappeared from the farming systems of the most economically advanced societies, and by this time were beginning to be supplanted in some developing countries as well.

Collecting becomes the concern of international organisations

FAO forms a Crop Genetic Resources Unit

In 1965, in response to the alarms which were being sounded by concerned individuals about the consequences to future crop improvement of the accelerating loss of genetic resources, the Food and Agriculture Organisation of the United Nations (FAO) established a small Crop Ecology and Genetic Resources Unit, to promote the collection and storage of seed material in secure conditions in what came to be known as genebanks. The resources available to the Unit were inadequate to deal with the scale of the problem, and in 1972 it called together a panel of experts to advise on the scope and urgency of the problem and to formulate policy on effective remedial action. The FAO Panel of Experts alerted governments and public opinion to the crisis which faced crop improvement, and to the consequences to world food production.

International concern leads to the founding of the IBPGR

The work of the FAO Panel led to the formation, in 1974, within the framework of the Consultative Group on International Agricultural Research (CGIAR), of the International Board for Plant Genetic Resources (IBPGR), an autonomous international scientific organisation to promote the collection, conservation, documentation, evaluation and use of crop plant genetic resources and to coordinate an international network of genetic resources centres.

Genetic resources conservation entered a new phase, characterised by a commitment to extensive international collaboration, to long-term security of collected material, and to the definition of scientifically based standards of description, documentation, storage and management and the promotion of their use in genebanks.

IBPGR, in its first ten years, stimulated and guided urgent rescue collecting on a world scale, of crop material under threat of erosion or extinction. Genetic resources collections increased dramatically in number and size. Estimates of the total numbers of accessions in the major collections for each species or species group are given in Table 4.1. The success and scale of this rescue operation inevitably revealed a range of scientific, technical and organisational problems which had to be solved. These problems included the devising and general adoption of standard descriptors for each crop – a common language of description understandable by all workers. It was necessary to determine optimum storage conditions for seed of each species. Seed samples were distributed to duplicate stores for security against loss due to civil unrest or natural disasters. The IBPGR also trained personnel and provided essential basic equipment to many developing countries.

As the numbers of accessions in collections increased, so did the amount of data. Without information on the species identity, source, seed numbers, viability and botanical and agricultural characteristics, collected material is of little use to plant breeders. Systems of hand-written records were unable to cope with the scale of data recording and analysis necessary for effective management of collections. It was soon apparent that the registration and processing of data was as big a task as the management of the seeds themselves and computerised data management systems were devised for use in and between genebanks.

What began over sixty years ago as the independent collecting of genetic resources by individuals or small teams, specifically to provide material for plant improvement programmes, has become a cooperative international rescue operation to preserve crop genetic diversity from total loss. Along with the collection successes, a steadily increasing number of organisational, technological and scientific advances are being made, often stimulated by the activities of the IBPGR.

It is important that crop genetic resources are not only safely preserved but that they are available to breeders for the improvement of varieties of our food and industrial crops. It is our purpose in the following chapters to discuss some of the principal issues in the conservation and use of crop genetic resources and, in so doing, we hope to reveal what crop genetic resource conservation means in present practice and to suggest what directions it may take in the future.

3

Genes to the rescue

Résumé

Plant breeding provides many examples of the use of genetic resources for the improvement of varieties of crop plants. These examples range from highly specific improvements to one major defect, such as susceptibility to a pest or disease, to all-round improvements in yield, agronomic characters and disease resistances and to changes in the form and structure of the plant.

Much attention has been given to the breeding of resistances to pests and diseases. This is partly because of the success of earlier breeding which had resulted in the adoption of genetically uniform varieties in place of heterogeneous landraces, with a consequent increase in disease and pest epidemics. This is reflected in the examples discussed in this chapter, all of which are outstanding successes in the use of crop genetic diversity. The examples are: resistance to potato cyst eelworm in potato and resistance to seven bacterial, fungal, viral and insect pathogens in rice; improvement of yield and nutritional value and disease and pest resistances in sweet potato; and the use of dwarfing genes to change the architecture of the wheat plant so that it can respond to higher fertiliser applications and give higher yields without collapsing.

Disease-resistance breeding differs from other breeding objectives in that the breeder must consider two genetic systems, those of the crop host and the pathogen, and their interaction. One consequence of the interaction is the ability of many pathogens to overcome resistance in the host plant so that many resistant varieties have a short effective life. Breeders have developed alternative strategies for the control of disease by genetic means, and all depend on the availability and use of genetic diversity.

In addition to the improvement of established crops there are dramatic examples of the use of genetic resources to create entirely new crops within recent times. We consider three examples: two industrial crops – sugar beet and rubber – and a new synthetic cereal – Triticale.

Genetic resources provide solutions to disease epidemics

In the last chapter we considered how landraces were formed and diversified in association with human societies as crops spread across the habitable regions of the globe and how, subsequently, humans took a positive hand in this adaptation process, first by simple selection and then by controlled breeding, to mould their crops more precisely to their needs. The deliberate adjustment of the characters of crop plants to suit the needs of the time was extremely successful and led to the supplanting of the variable landraces by more uniform varieties. The new varieties spread over wide areas and led to the replacement of diversity by uniformity. Some time after this trend was established, it became apparent that it carried unforeseen penalties. There seemed to be a more frequent occurrence of disease epidemics and, in consequence, greater variability in yields.

Crop plants are subject to attack by a wide range of pathogenic organisms including fungi, viruses, bacteria, insects and nematodes. Each crop plant has its own spectrum of possible pathogens. Likewise, each pathogen is limited in the range of host plants which it can attack. For example, the fungus causing black stem rust of wheat cannot cause the same or any other disease in sugar beet or even in other cereals such as rice or sorghum. Some parasitic fungi have wider host ranges, but all have strict limits. Within this parasitic-species to host-species relationship, there is a finer degree of mutual genetic adjustment in which a strain or race of black stem rust may be able successfully to infect some varieties of wheat but not others. The establishment of an infection on a host plant by a parasite depends on the interaction of resistance genes in the host and correspond-ing genes for pathogenicity in the parasite.

Landraces, characteristically, are genetically heterogeneous. This is manifested, in a landrace of wheat, by obvious diversity in the appearance of the plants. There can be much variation in length of straw, architecture of the ear and colour of the grain. This diversity extends to characters which cannot be readily seen, such as in their genes for resistance and susceptibility to their various pathogens. Landraces present a genetically complex target to the pathogens, thereby moderating the more extreme expressions of disease.

In contrast, a modern homogeneous, genetically uniform wheat crop presents, in each plant, the same gene for resistance or susceptibility to a fungus. If the fungus possesses a gene for pathogenicity capable of establishing infection on the particular wheat variety, then all plants are liable to be infected and an epidemic can ensue. The consequences of this recipe for disaster were quick to show following the large-scale adoption of high-yielding varieties. Epidemic outbreaks of disease, though not unknown in landraces, became more frequent and losses more severe.

Not surprisingly, breeding for resistance to disease became a major objective of many breeding programmes. When resistance genes were not available in their current breeding material, breeders had to turn to landraces and primitive varieties as possible sources of the genetic diversity which they needed. If these sources failed, they had to go further afield, to related wild species. In other words, they explored the genetic resources of the crop, turning first to the more closely related material and moving on to more distant relatives as necessary. Hybrids between modern varieties and wild species often present problems of sterility and agricultural acceptability which may take years of breeding to overcome, and so breeders prefer to use genes in the agriculturally adapted genetic backgrounds of landraces and old varieties, if they can be found.

Disease resistance is not the only type of breeding problem for which genetic resources are necessary, but it provides us with some of the most dramatic examples of the way in which genes from primitive varieties and wild species have been used to improve crop performance. In this chapter we present three examples concerned with disease and pest resistance and one concerned with the modification of the form of the plant. In all cases, the use of genetic resources led to large gains in crop production.

Genes control potato cyst eelworm

Eelworm is a major potato pest

The potato cyst eelworm is a tiny soil-inhabiting nematode which invades the roots of potato plants and, having fed off its host, emerges from the root to complete its life cycle by turning itself into a small rounded cyst, about the size of a pin-head, containing a number of larvae. The cyst remains intact in the soil and the larvae retain their viability until the next potato crop, when an exudate from the roots of the host plants stimulates the larvae in the cyst to develop into the next generation of eelworms, which repeats the cycle.

The growing of successive crops of potatoes on the same land leads to a rapid build-up of the eelworm population. Damage to the host plants can be severe when they are invaded by large numbers of the eelworms and reduction in the yield of tubers can amount to total crop failure. Progressive crop losses became a common experience in the traditional potato-growing areas of many countries and some farmers were forced to abandon potatoes and to turn to alternative crops. Potato production began to move to soil types less suited to the crop but which were eelworm free. By the late 1930s, potato cyst eelworm was becoming a major limiting factor to potato production in many European countries, and no practicable means of chemical control was available.

A resistance gene gives protection

Conrad Ellenby, a zoologist in the University of Newcastle upon Tyne, studying the biology of the eelworm (*Heterodera rostochiensis*), screened the Empire – now Commonwealth – Potato Collection (CPC) to determine the reaction of the accessions to the eelworm. The CPC was one of the earliest comprehensive genetic resources collections and was assembled to provide much-needed genetic diversity for potato-breeding programmes. Between 1941 and 1952, Ellenby systematically tested more than 1200 accessions belonging to more than sixty cultivated and wild *Solanum* species collected from South America. He grew the material on public allotments where potatoes had been grown, probably every year, since the allotments were started in World War I; a simple, severe and effective test (Ellenby, 1952). He found that plants of six accessions were resistant – they had few or no cysts on their roots. Four of these accessions belonged to *Solanum tuberosum* sub-species *andigenum*, which is probably the immediate ancestor of the European potato (*S. tuberosum*). The close relationship was important because hybrids between the two cultivated forms were easily produced and were fertile, and breeders were able to use the subsp. *andigenum* material as parents in breeding for resistance.

In the same year, two Dutch workers following Ellenby's lead, reported the first evidence that resistance was genetically controlled (Toxopeus and Huijsman, 1952). Subsequently, control was shown to reside in a single gene (designated H1), and single-gene control favours rapid progress in breeding. By the late 1960s the first of a succession of resistant varieties appeared in Britain and in continental Europe enabling farmers to produce high-yielding crops on land which previously produced failures. The H1 gene has had a profound effect on potato production in many countries where the potato is a major crop. There is no environmental cost of this method of disease control and potato producers are in no doubt about the value of this genetic resource in solving a major problem.

There is an additional benefit from the action of the H1 gene. Its action is to interfere with the reproductive life cycle of the eelworm. Larvae are stimulated to hatch and invade the host roots, but fail to form cysts. Consequently, while the host stimulates the larvae to develop, the H1 gene prevents them leaving any progeny and so achieves a cleaning action on the soil.

New eelworm races required new resistance genes

It transpired that the potato eelworm populations were genetically diverse for pathenogenicity and that the H1 gene conferred resistance to only part of this diversity. This is a common experience in disease-resistance breeding. Discrimination between different genetic forms of the pathogen, whether nematode, fungus or virus, is only possible when genes for resistance begin to operate in the host. Fortunately, other accessions in the CPC were identifed which possessed resistance to the other races and other species of eelworms, and so breeding was able to continue.

Rice: high-yielding varieties bring problems

The IRRI transforms the production of Asian rice

The International Rice Research Institute (IRRI), was established in 1960 as an international agricultural research centre, later to become one of the Centers of the CGIAR. It appointed a multi-disciplinary team of scientific specialists which has produced a series of progressively improved varieties which dominate rice production in Asia. Their first major success, with which the 'green revolution' in rice began, was the variety IR8, released in 1966. It was short-strawed, high-yielding and adapted to a wide range of environments. Typical yields of tropical rices at that time were 2.0–3.5 tonnes/hectare when given 30–40 kilograms (kg) of nitrogen/ha – more nitrogen resulted in stem collapse and decreased yields. IR8 produced record yields in experimental plots in 1966–68 of up to 11 t/ha in response to nitrogen levels up to 150 kg/ha (Chang, 1976). Not surprisingly, its cultivation spread with great speed, and it replaced large numbers of landraces and old cultivars, of which it was said that there were more than 5000 in India alone (Purseglove, 1985a). Serious constraints on IR8's high-yield potential quickly became apparent. It was susceptible to seven

Table 3.1. *Progressive development by IRRI of some pest and disease-resistant varieties of rice. (Extracted from Plucknett et al., 1987.)*

Variety	Year released	Brown plant hopper races			Green leaf hopper	Stem borer	Grassy stunt	Tungro	Blast	Bacterial blight
		1	2	3						
IR2	1966	S	S	S	MR	S	S	S	S	S
IR20	1969	S	S	S	R	MR	S	MR	MR	R
IR26	1973	R	S	R	R	MR	S	MR	MR	R
IR28	1975	R	S	R	R	MR	R	R	R	R
IR36	1976	R	R	S	R	MR	R	MR	R	R
IR38	1976	R	R	R	R	MR	R	R	R	R
IR46	1978	R	S	R	R	MR	S	R	R	R
IR56	1982	R	R	R	R	MR	R	R	R	R

Notes:
S = susceptible, MR = moderately resistant, R = resistant

serious diseases and pests, namely, grassy stunt virus, tungro virus, blast (a fungal disease), bacterial blight and, among the insects, stem borer, green leaf hopper and three types of brown plant hopper (Plucknett *et al.*, 1987).

IR8 was followed by a series of IR varieties which progressively overcame the deficiencies of IR8 and, in 1982, IR36 alone was grown on eleven million hectares, the most widely planted rice variety in history. Its use has now spread to some of the rice-growing areas of Africa.

Displaced landraces provide needed resistance genes

From the outset, IRRI set about the systematic collection of Asian landraces and old varieties and in the first three years had acquired more than 10 000 samples. By 1989, the figure had risen to 83 000, out of an estimated total of about 100 000 rices which had existed in Asia (Chang, 1989) before they were largely swept away by the new varieties.

The collected material was subjected to intensive examination for characters of importance including resistance to the major disease organisms. As resistance genes were found they were incorporated into the high-yielding genetic background of IR8 and its derivatives. IR26, released in 1973, possessed resistance or partial resistance to all the listed pathogens except grassy stunt virus and one type of Brown Plant Hopper (see Table 3.1). In 1976, IR36 was released with additional resistance to grassy stunt. This resistance was found after screening 17 000 accessions of cultivated Asian rice, *Oryza sativa*, from the IRRI collection along with more than a hundred wild species. In one accession of an annual wild rice (*O. nivara* – the immediate progenitor of *O. sativa*) – from Uttar Pradesh, three plants out of thirty tested contained the resistance gene (Chang, 1989). This is a vivid illustration of the importance of maintaining representation of the full range of the genetic diversity of wild species. Finally, in 1982, the breeding sequence was completed with the release of IR56, which possessed resistances to eight of the pathogens and partial resistance to stem borer (Plucknett *et al.*, 1987). By any standards, this must be regarded as an astonishing achievement, particularly as the expression of other characters such as yield, grain quality and tolerance to adverse soil conditions were maintained or improved.

Losses due to brown plant hoppers were particularly serious because, in addition to the direct damage they caused by sucking the sap of plants, they could simultaneously transmit grassy stunt virus to the plant on which they were feeding. The resistances to the pest and to the virus are under the control of different genes which come from different sources. The assembly of widely dispersed genetic material into IR36 entailed an elaborate crossing and inter-crossing programme with selection among the progeny based on systematic screening tests for reaction to each disease and pest. The presence of a resistance gene can only be detected by challenging the plant with the pathogen or pest and observing the result. The design of suitable screening procedures and their large-scale application is an integral part of breeding for resistance.

There can be no end to breeding better varieties

The rice story does not stop here. The successful assembly of genetic resistances to the array of pathogens, is the end of the beginning rather than the beginning of the end of disease-resistance breeding in rice. Plant pathogens present a constantly moving target. Other diseases appear and become important limitations to yield. Ragged stunt was previously of little significance but is now of mounting economic importance. Frequently, new mutant forms of common disease organisms appear with different genetic capabilities for infecting the host, enabling the pathogen to by-pass those resistance genes incorporated into the first wave of resistant varieties. This sequence of action and reaction has been a common feature of resistance breeding in other crops; for example, in resistance to fungal rusts in wheat and to late blight in potatoes. The same sequence appears to be occurring now in rice. There is evidence of the widespread occurrence of new strains of grassy stunt which can attack IR36. Therefore, despite its phenomenal success, with yields in farmers' fields of between 4 and 6 t/ha, IR36 can only be expected to have a life of say ten to fifteen years before it too will need to be replaced by newer varieties with new disease resistances and even better agronomic characterics. Experience shows that most, although not all, resistant varieties have a transient life. There can be no end to the continuing struggle to cope with the generation of new virulent forms of pathogens. Success in this enterprise depends on breeders having access to adequate genetic variability of their crop plant.

The aim of this type of resistance breeding is always to keep one step ahead of the pathogen, and intelligent prediction, based on the constant monitoring of pathogen race populations, of what resistances will be needed in the future, is a necessary component of resistance breeding.

Wild species are important in sweet-potato breeding

A subsistence crop awaiting improvement

Sweet potato (*Ipomoea batatas*) is a starchy root vegetable of the tropics and sub-tropics and provides a striking example of the potential for crop improvement which lies in the genetic diversity of wild species. In contrast to our other examples, potato, rice and wheat, it is a crop which, until recently, has received relatively little attention from plant breeders.

Its origins lie in Central America but exactly where is not known. According to Purseglove (1982a), it was grown in pre-Columbian times in Mexico, and other parts of Central and South America. It was transported from these areas to Polynesia and New Zealand. At that time it was unknown in Africa, Asia and Europe, but today it is widespread, not only in the tropics and sub-tropics, but also in some warm temperate regions. Annual world production is over a hundred million tonnes, much of it in Central America, Africa, south Asia and the Pacific. In some developed countries, such as the USA, Japan and New Zealand, it is grown for sale as a cash crop. Elsewhere, in areas where major population increases are predicted, it is an important staple in the diet of subsistence farmers. It is in this connection that its improvement is, perhaps, of most significance, especially because large gains can usually be made in the early stages of breeding and selection from unimproved material.

Wild relatives provide pest and disease resistances

Until recent times, sweet-potato breeding was mainly confined to inter-crossing between cultivated clones – the crop is normally propagated vegetatively by cuttings or by tubers. More recently, non-tubering wild species have been used as donors of genes for disease and pest resistance (Iwanaga, 1987). Sweet potato has no known wild tuber-forming ancestor to which breeders would naturally have turned first as a source of genetic variation. The evolutionary pathway of *I. batatas* is unclear, as are the relationships among species in the genus, but interspecific hybrids can be produced.

Promising exploratory work has been carried out in Japan and China on hybrids between local cultivated clones and two wild species, *I. trifida* and *I. littoralis*. Japanese workers identified a form of *I. trifida* with resistance to two soil pests, root-knot nematode and root-lesion nematode. These resistances were successfully combined, by suitable crossing procedures, with high yield and high starch content. The outcome in 1975 was the successful commercial variety Minamiyutaka, now widely grown in Japan. Studies on the potential of other hybrids between *I. batatas* and *I. trifida* were made in China using local varieties of *batatas* as the cultivated parent. The characteristics of the hybrids varied widely, but some produced high yields of tubers with exceptionally high dry-matter content (around 40 per cent), when compared to the cultivated parents, where the norm is about 30 per cent.

In a collaborative project involving scientists from Japan, the Asian Vegetable Research and Development Center (AVRDC), and the Centro Internacional de la Papa (CIP), the characteristics of other hybrids derived from the same two species were analysed for yield of tubers, dry-matter contents and for resistance to the sweet-potato weevil (*Cyclas formicarius*), the major pest in most countries (Iwanaga, 1987). More than 600 hybrids were analysed. Some had combinations of acceptable tuber yield, exceptionally high dry-matter content (up to 45 per cent), and high protein contents (up to 3 per cent). There was evidence of high levels of weevil resistance in some of the high-yielding clones.

Similarly, in hybrids between Chinese *I. batatas* varieties and the wild species *I. littoralis*, tuber yields and dry-matter contents were in some case much higher then the local cultivated parent, and 31 per cent of the hybrids were highly resistant to the important black rot disease (*Ceratocystis fimbriata*). The evidence from these exploratory studies on the potential of wild germplasm for sweet-potato improvement is extremely encouraging. The proportion of the wild gene-pool which has been sampled so far is miniscule, and yet resistance to a major disease and a major pest and genetic variation for the increase of tuber and starch yields have been uncovered.

The potential of the gene-pool cannot be evaluated without breeding

The wild species of sweet potato do not produce tubers at all. Yet most of the breeding objectives are related to characters of the tuber (yield, starch content, flavour, protein content, earliness of tubering, storage characteristics and weevil resistance). Therefore the potential value of wild species cannot be assessed directly from their own characteristics but only through the production and analysis of hybrids with *I. batatas*. However, present evidence indicates that the formation of storage tubers is controlled by a few genes, which facilitates breeding and selection.

Work on the improvement of sweet potato is at an early stage and until recently was largely confined to two developed countries, the USA and Japan, but enough has been done to indicate the great potential of wild species. Now, with major breeding inputs from two International Agricultural Research Centers, AVRDC in Taiwan and CIP in Peru, in collaboration with national programmes in China, Japan and the USA, on the utilisation of wild species in sweet-potato improvement, we may expect spectacular improvements in a short time. Collaborative studies are in hand on the limitations to production in developing countries. Pests, diseases (including important diseases due to viruses), and other environmental factors such as adverse soil conditions, are being defined as breeding goals. Breeders are exploring additional accessions of *I. littoralis* and *I. trifida* and of the other wild species capable of hybridisation with *I. batatas*, to determine the most promising parental material (Iwanaga, 1987).

Shorter wheats: bigger yields

Another of the great success stories of modern plant breeding is the development of dwarf and semi-dwarf, high-yielding varieties of bread wheat or, more correctly, varieties of high-yield potential, for high yields in the field require improved agronomy as well as improved genetic constitution of the variety.

Less straw and more grain means a higher harvest index

Careful analysis of the stem, leaf and grain production of twelve winter wheat varieties, mostly modern but one, Little Joss, introduced into agriculture in 1908, has revealed the interesting point that the total amount of above-ground dry-matter production has not changed in the varieties which span this eighty-year period. However Hobbit, a release of the 1970s, outyielded Little Joss by 40 per cent in grain production (Austin *et al.*, 1980). What has changed is the partitioning of the products of photosynthesis – the photosynthates – on which growth processes depend. In modern varieties, more photosynthate goes into ear and grain production and less into the production of stem; they have a greater harvest index. The leaf area – which is related to the amount of photosynthate – of the old and new varieties is about the same. This means that there was an improvement in the harvest index which had been achieved in incremental steps during eighty years of interbreeding among the adapted varieties, a process repeated in many wheat-producing countries. By the mid-1970s, it had reached a point where further progress was increasingly difficult to achieve.

Norin 10 dwarfing genes improve world wheat production

The slow incremental progress in raising wheat yields suddenly changed up a gear with the identification of dwarfing genes of large effect. These genes had the specific function of controlling the length of the stem between successive leaves. The incorporation of dwarfing genes into modern varieties has had a dramatic effect on wheat production world-wide, and led to the introduction of wheat varieties which, together with the series of rice varieties from IRRI discussed above, gave rise to the sensational increase in grain yields – particularly in developing countries – which we know as the 'green revolution'.

The effect of dwarfing genes is to reduce the length of stem without reducing the leaf area of the plant. They also cause some reduction in grain size, but this is compensated by higher grain numbers per ear. The degree of expression of these characters varies according to the rest of the genetic complement into which they are introduced. The general consequence of their use has been the production of varieties with larger ears carried on shorter, stiffer straw, which is less liable to collapse or 'lodge' before harvest. Lodging can lead to serious loss of grain at harvest. Stiff-strawed dwarf varieties have another more positive merit in their ability to respond with increased grain yields, but without lodging, to higher levels of fertiliser application than is possible with longer strawed varieties. The yield advantage of the dwarf or semi-dwarf wheats is therefore a composite of these attributes.

About a dozen major dwarfing genes from different sources have been identified in the past fifteen years, but most of the current short-strawed varieties owe their semi-dwarfness to two – Rht1 and Rht2.

Japanese breeders were the first to utilise the dwarfing genes in wheat which they doubtless obtained from indigenous landraces. Two general lines of development can be traced. The first began with the old Japanese variety, Akagomugi, bred in the latter part of the 19th century. Akagomugi was introduced as a parent into Italian wheat breeding in 1911, and gave rise to a series of successful varieties. These were in turn used as parents for introducing strong straw into other European programmes and also that of the Rockefeller International Wheat Improvement Project in Mexico (later to become the Centro Internacional de Mejoramento de Maiz y Trigo (CIMMYT) (Gale and Youssefian, 1985).

The second introduction of the dwarfing genes into modern wheats began with another old Japanese variety, Daruma (see Fig. 3.1). Its derivative, Shirodaruma, was crossed in Japan in 1917 with the North American variety Fultz. The product, Fultz–Daruma, was subsequently

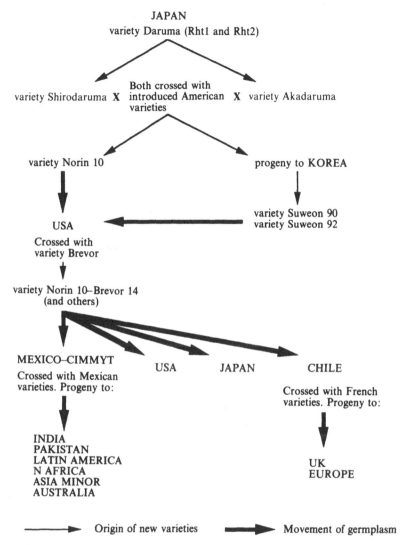

Fig. 3.1 Scheme showing origin of the Norin 10 dwarfing genes Rht1 and Rht2 and their dispersion through world wheat-breeding programmes. (Based on Gale and Youseffian, 1985.)

crossed with the American variety Turkey Red from which came the most important outcome of this work, namely Norin 10, released in 1935 (Lupton, 1987). Norin 10 made no major impact on Japanese agriculture but was used in 1948 as a parent in a breeding programme at Washington State University, which aimed to produce varieties able to respond, without lodging, to heavy applications of nitrogenous fertilisers. The variety Norin

10 – Brevor 14 was the first product of this programme to reach the farmer. More importantly, it was the vehicle by which the two Norin 10 dwarfing genes were to reach other breeding programmes in the USA, in Mexico and – through CIMMYT's international programme – most wheat-producing countries and regions including India, Pakistan, Latin America, Australia, Canada, Argentina, Asia Minor, China, Europe and, of course, Japan. The spread of these genes from the Washington State programme has been astonishing in its scope and speed. Their impact on world wheat production is incalculable and still continues. It has been estimated that further increases in grain yield by genetic manipulation of the harvest index may raise it from its present level of nearly 50 per cent to nearer 60 per cent, which is thought to be about the upper limit. There is clearly a point beyond which the straw may not be further shortened without reducing the functional efficiency of the leaves to intercept light energy. When this point is reached, further progress is likely to depend on the detection and exploitation of genetic variation for increased efficiency in converting light energy into chemical energy and, for this new goal, breeders will again have to turn to the genetic resources of the crop.

The use of dwarfing genes has disadvantages

Two additional points need to be made. Yield gains achieved by means of the dwarfing genes have disadvantages. Evidence is accumulating that high-yielding dwarf and semi-dwarf wheats are less stable in their grain yields from year to year than their non-dwarf predecessors (Gale and Youssefian, 1985). They may be more sensitive to fluctuations in annual weather cycles, and seem to be more prone to disease epidemics. This is a matter of some importance because a high proportion of world wheat production is now dependent on varieties carrying the Norin 10 genes, and stability and predictability of yields are of great importance in the planning of food production by national and international policy-makers. Indeed, stability of yield is becoming an important breeding objective in many crop improvement programmes.

The second point concerns the validity of the strategy of breeding varieties which require high inputs of artificial fertilisers for the realisation of their maximum yield potential. This is a matter receiving much attention at present. Should breeding strategy for wheat change toward maximising yields from low-input systems, breeders will once again have to turn to the stored genetic diversity of landraces and primitive varieties to achieve their new breeding objectives.

World wheat production has benefited from the international exchange of genetic material

The history of the use of the dwarfing genes in wheat provides an excellent example of the benefits of unrestricted exchange of genetic resources. American wheat varieties were used in Japan in the breeding of Norin 10, which was then used in the USA to breed the dwarf and semi-dwarf varieties used subsequently as parents in Japan where the genes originated, and in most other wheat-producing countries, including developing countries (Fig. 3.1). The benefits of this free exchange have been enjoyed by wheat growers and consumers world-wide.

Host–parasite relations: an evolutionary battleground

Disease levels depend on the interaction of two genetic systems

The driving force behind the continuing interactive struggle between the crop host and its pathogen comes from man's disturbance of the equilibrium which exists between the populations of the two organisms. When a genetically uniform variety such as IR36 is widely grown as pure stands, it applies a severe selection pressure against those genetic types (genotypes) of the parasite to which it is resistant, but in favour of any to which it is susceptible. These latter, if initially rare in the pathogen population, are able – when freed of competition from the previously dominant strains of the pathogen – to establish infections, to multiply and to succeed as the new dominant pathogenic races. The greater the area on which the resistant variety is grown, the greater the probability that a hitherto rare pathogenic mutant will be selectively multiplied in this way.

The rate of spread of new strains of pathogens varies from one parasite to another and relates to the method of dissemination of the species. Wind-borne spores of fungal parasites, such as the cereal rusts and mildews and potato blight, can spread new races over great distances with alarming speed, as can mutants of viruses transmitted by insect vectors which have large dispersion ranges. Soil nematodes, on the other hand, are carried passively in soil, and depend usually on the movement by man of underground plant parts such as roots and tubers with soil adhering to them. In this way both the crop and its pathogen are propagated together. The biology of the pathogen clearly affects the rate at which epidemics build up and the response time available to the breeder in which to counter them.

Southern corn blight – an epidemic caused by plant breeding

A dramatic example of epidemic disease in modern agriculture is provided by the case of southern corn blight in the USA in the early 1970s. Its appearance was sudden, its spread, by means of air-borne spores, extremely rapid, and its effects disastrous. Yield reductions of up to 40 per cent and a depression of the national yield in 1970 by about 15 per cent, were estimated. The cost to the US industry amounted to hundreds of millions of dollars (Plucknett *et al.*, 1987). Losses in subsequent years were minimised by the high degree of organisation and flexibility in the US plant-breeding and seed-production industries which enabled the rapid substitution on the farm of varieties resistant to the fungus.

The problem arose because, by the late 1960s, US corn production was based on hybrid varieties obtained from female parents, all of which carried a factor for male sterility in the cytoplasm. This factor, a genetic emasculator, was used to ensure that the seed produced by these plants was the result of a cross-fertilisation and was hybrid.

Unfortunately, the male-sterility factor also conferred susceptibility to a new race of the southern corn blight fungus. This race appeared in the late 1960s and had a clear run through the US hybrid corn crop which was almost uniformly susceptible. However, the epidemic had one beneficial consequence. It focussed attention on the hazards inherent in crop genetic uniformity and on the importance of collecting, conserving and using genetic resources.

Breeders have alternative strategies against disease

The foregoing discussion on breeding for disease resistance, relates to resistances which are controlled by single genes of major effect. This type of genetic control is common and is favoured by breeders because of the relative simplicity and speed with which it can be transferred from parents to progeny in a breeding programme. It is usually qualitative in its effect, that is, a plant is either resistant or susceptible. This type of interaction is often described as 'vertical resistance'.

There are many and perhaps more numerous examples where resistance in the host is expressed in a quantitative manner, that is to say, resistance is a matter of degree. This type of resistance is described as 'horizontal resistance'. It is less likely to be circumvented by the pathogen – it is said to be more durable – and it is usually under more complex genetic control, perhaps by several or many genes. The precise interactions between single genes in host and pathogen, characteristic of vertical resistance, do not apply here. Instead, there is partial resistance irrespective of pathogen race.

The claimed durability of horizontal resistances make them at first sight an attractive alternative strategy for the genetic control of disease epidemics. However, the practical problems of transferring the requisite genetic factors from a primitive horizontally resistant parent, and at the same time eliminating from the progeny other unwanted genetic material derived from that parent, can be formidable. Recombination between the genomes of the two parents takes place in the hybrid and in its subsequent progeny, and is a process out of the control of breeders; they can only select among the combinations presented to them. The usual experience is that offspring with acceptable levels of resistance usually have unacceptable expressions of other characters of importance.

This point is illustrated by our earlier example of breeding for eelworm resistance in potatoes. The vertical resistance conferred by the H1 gene derived from a primitive but cultivated variety, was quickly transferred into adapted genetic backgrounds and into successful new varieties. In marked contrast, later attempts to breed varieties combining a high level of race-non-specific resistance from the wild species, *Solanum verneii*, together with acceptable expressions of the many other characters of importance to growers and consumers, proved to be a long and difficult process. These difficulties have to be balanced against the benefits of durability and against the apparently endless requirement for new varieties when depending on novel vertical resistance genes. Increasingly, breeders are turning to horizontal resistance when it appears to be a feasible option.

The third strategy for disease control attempts to combine the benefits of genetic diversity for resistance genes characteristically found in landraces, with the advantages of uniformity and high yield, characteristic of modern high-yielding varieties. This is done by creating varieties from artificial mixtures of say six or eight selections which differ only in the vertical resistance genes which they carry. These are known as multiline varieties. In theory, all components will be susceptible to one or more of the races of the pathogen, but the specificities of each will differ. Collectively, the mixture will be constructed so that it possesses resistance genes to all the common races of the pathogen and each plant will act as a trap for spores to which it is resistant and so provide protection to other genotypes in the mixture. The effect is to delay the development of the disease in relation to the growth and maturation of the crop and hence to reduce the amount of damage, but not to eliminate it. Selection pressures applied to the pathogen population will be less than in a variety with genetically uniform resistance, and therefore the probability of genetic changes in the pathogenicity of the parasite population will be less.

The breeder must maintain a supply of component lines possessing different resistance genes to those currently in use and substitute these in the multiline mixture in response to changes which may occur in the race composition of the pathogen population.

This option of ringing the changes on component lines, relieves the breeder of the apparently endless commitment to respond by breeding varieties with new resistances. However, it has its own costs in the creation, multiplication and maintenance of reserve lines, a complex and expensive operation. Component lines, in theory, differ only in the resistance genes which they carry, but in practice they not infrequently differ in other respects also, and lack of uniformity can cause problems in the growing and harvesting of the crop, and in the marketing and use of the grain.

Genetic control of crop diseases will become even more important in future

The pursuit of disease control by genetic means is likely to remain a major objective in most crop-breeding programmes, because of its cost-effectiveness and, increasingly, because it is a form of biological control and is free from the deleterious environmental side-effects associated with chemical control methods. While recognising the necessity of using chemicals, as an interim measure, to control diseases, it seems certain that we shall look more and more for genetic solutions to disease problems for there is accumulating evidence of environmental damage resulting from chemical sprays. The strategies adopted will depend on the circumstances of the crop and on the genetic resistance system available. The important point is that all three options are dependent on the genetic resources in the crop gene-pool. The case for their collection and conservation to meet present and future requirements is overwhelming.

Genetic resources are the raw material for the creation of new crops

New crops are designed to meet particular needs

A small number of new crops have been created in recent times. Their development follows the same general sequence as the evolution of the 'old' crops discussed above, but differs in a number of important ways. They are, from the outset, the product of directional breeding and selection to meet a defined specification. They often have a relatively narrow genetic base because the normal evolutionary process of the 'old' crops has been condensed from several thousand into less than two-hundred years and the outcome has been relatively few cultivated varieties produced without the prior evolution of thousands of landraces.

Sugar beet, Beta vulgaris

Beets in their various forms are derived from the wild *Beta maritima*, a variable species which is locally common around the shorelines of the Mediterranean and of western Europe, although in the past two decades it has become rare in many areas. The use of beet as a leafy vegetable goes back to prehistoric times, and the development of leaf beets as we know them today probably began in southern Italy. Chards, or Swiss chards, is a specialised form of leaf beet grown for its fleshy white mid-ribs. Two other ancient cultigens are grown for the food reserves in their swollen roots. Red beet (beetroot) probably evolved in the Caucasus and Macedonia. It was used by the Romans and had spread to central and northern Europe by the Middle Ages. By the middle of the 18th century, another swollen root form – fodder beet – had evolved in central Europe where it was widely grown as stock feed and from where it spread to France and northern Europe. The roots of the landrace fodder beets had sugar contents of about 6 per cent (Campbell, 1976).

The three forms evolved in different places but all had their origin in the wild gene-pool of *B. maritima* from the eastern Mediterranean (Ford-Lloyd and Williams, 1975).

According to Campbell, the stimulus for the development of the sugar-beet crop came from two events. The German chemist Achard, under the patronage of the King of Prussia, devised a process for extracting crystalline sugar from root sap, and the first factory to manufacture sugar from fodder beet was built in Silesia in 1801. The second stimulus came from the blockage by the British navy of French ports, which disrupted the supply of cane sugar from the West Indies, and prompted Napoleon, in 1811, to decree that beet should be grown to provide an alternative source of sugar and that the culture of the crop should receive serious study. So the development of beet as a specialised sugar-producing crop began; selection was for increased sugar content of the root sap, for yield of roots and against flowering in the year of sowing – 'bolting' – which significantly reduces the yield of sugar. Simultaneously, there was a progressive increase in the efficiency of the extraction and refining processes which contributed to the success of the crop.

The modern sugar-beet plant differs markedly from the fodder beets of the early 19th century. Sugar contents now approach 20 per cent, most modern varieties are triploids (they are more productive than diploids), and the agricultural 'seed', which is a hard fruit containing several seeds in old varieties, has been changed to the single seeded – 'monogerm' – condition. This obviates the need for the expensive singling of groups of seedlings in the field after germination. Sugar beet is now a major industrial crop grown on some seven million hectares in Europe and the former USSR, and on another million hectares elsewhere in Asia.

Triticale, Triticosecale

Triticale is a synthetic cereal derived from hybridisation between wheat (*Triticum*) and rye (*Secale*). Natural hybrids had been the subject of study since the end of the 19th century, but serious work on the development of Triticale as a crop plant did not begin until the 1930s in Sweden and the 1940s in the USSR. In the early years the intention was to produce a cereal with the hardiness of rye and the yield and grain characters of wheat which would extend the area of high-value cereal production. Subsequently, breeding objectives have been modified as the potential of Triticale has become clearer.

The early Triticales possessed eight genomes, AABBDDRR; the six of bread wheat, AABBDD, and the two of rye, RR. They were characterised by variable but generally low fertility, and a tendency to produce shrivelled grain; serious handicaps in a crop intended for grain production. Interest began to languish until it was found that hybrids with six genomes (AABBRR, derived from tetraploid Durum (or macaroni) wheat, AABB, and rye), possessed higher and more stable fertility. This led to large-scale breeding programmes, first in Canada in 1954, then at CIMMYT in Mexico in 1965 and then jointly by the two groups (Larter, 1976). The release of commercial varieties began in the 1970s in Canada and Mexico and later in other countries.

Much of the interest in Triticale results from its high level of resistance to most of the diseases which attack wheat. However, as the areas sown to Triticale increase, so too do the selection pressures applied to the populations of the disease organisms. Predictably this will lead to the selective multiplication of pathogenic races capable of establishing infections.

A more permanent attribute of Triticales is their tolerance of soil conditions which prevent the growing of wheat, for example high acidity. They tolerate soils with high aluminium and low copper contents which are toxic to wheat, and can grow on poor sandy soils with a low water-holding capacity.

By 1986, Triticale was grown on more than 750 000 hectares world-wide, in Australia, France, the USA, and Poland. The area in Poland was expected to reach one million hectares by 1990, and to replace rye on those light soils where wheat cannot be grown economically. In western European countries its future seems to be as a replacement for winter barley (Gregory, 1987). In recent years, much attention has been given to widening the genetic diversity of the Triticale breeding pool by synthesising new hybrids from wheats and ryes from different parts of the world with the purpose of increasing the range of habitats in which Triticale may be grown successfully. This procedure is essentially the same as the sequence of events which gave rise to the many landraces of wheat and rye, but it is on a smaller scale and at a much faster rate.

The best Triticales have grain yields equal to modern high-yielding wheat varieties. With modified milling techniques they give similar yields of flour but, despite modifications to bread-making methods, Triticale bread is still inferior to wheaten bread (Gregory, 1987).

The ultimate economic success of Triticale as a cash crop is, as yet, uncertain despite its considerable potential. It may well achieve a comparative advantage over cereal alternatives in harsher climates where wheat is reaching the limits of its cultivation. When grown for home consumption by small subsistence farmers, its success will depend on its palatability and suitability for local traditional cooking methods as well as on its productivity.

The development of Triticale was the first attempt to exploit a possibility which has become more apparent in recent years, that wheat, barley and rye have one gene pool which includes species of all three crops and many related wild grasses (known collectively as the Triticeae). These species have distributions which together encircle the northern hemisphere, and some extend down the Andes. With the development of techniques of gene transfer from species to species and from genus to genus, which do not depend on sexual reproduction, this huge gene-pool is becoming more accessible and it seems highly likely that future breeding in these crops will involve species which, not long ago, were thought to be too remote to be of use.

Rubber, Hevea brasiliensis

Rubber has an amazing history of domestication and development from a wild species directly to a major crop and industrial commodity in the space of a hundred years.

There are nine recognised species of *Hevea*, all native to South America in the Guianas, the valleys of the Amazon and Orinoco and in the Matto Grosso. The story of rubber depends, essentially, on one species. *H. brasiliensis*, although others do yield commercially acceptable latex. The home of *H. brasiliensis* is the tropical rain-forest of the southern part of the Amazon basin.

In 1839, the process of vulcanisation of rubber was developed, and this event of great industrial significance led to the extensive exploitation of the wild trees of the Amazon. Supplies from this source were soon inadequate to meet world demand and Sir Clements Markham and Sir Joseph Hooker (the latter was Director of the Botanic Gardens, Kew) arranged for the collection of seed which was sent to Kew in 1873. From this seed twelve plants were sent via Calcutta to Sikkim in India, where they apparently failed. A second consignment of seed sent direct to India in 1875 failed to germinate.

Hooker then commissioned H.A. Wickham, who was in Santerem, on the Amazon, to collect a larger quantity of seed. He collected 70 000 seeds, all of *H. brasiliensis* and from an ecotype of high-yield and high-quality latex. Wickham chartered a ship to carry the seeds without delay to Liverpool, whence they were taken by special train to Kew, arriving in June 1876. Less than 4 per cent germinated, but 2800 seedlings were raised.

The Government of India was involved in this venture, and had financed the seed collection, but at this point lost interest because, apparently, of commercial uncertainties about the future of a rubber crop in India. In August 1876, the British Colonial Office sent nearly two-thousand seedlings in specially constructed glass cases, to Ceylon (Sri Lanka). Kew also sent another twenty-two plants to the Singapore Botanic Gardens in 1877; nine were planted there and another nine in the garden of the residency in Perak.

So, by the end of 1877, young trees were established in Ceylon, Perak and Singapore, and by the early 1880s all were fruiting. From the trees in Ceylon, 20 000 seeds were harvested in 1888, and some of these were sent to Singapore. By the mid-1880s, therefore, material derived from the original

Wickham collection was plentiful in south Asia, but commercial interest was nil.

At this point, H.N. Ridley was appointed as the first scientific director of the Singapore Botanic Garden and he set about the commercial exploitation of the rubber genetic resources which he found in his charge. There were nine trees of the original introduction, twenty-one five-year-old trees, and 1000 seedlings. He laid the foundations of a new industry by linking the biological properties of the species with the agronomic techniques required by the crop, to an industrial and commercial opportunity. Among his achievements were the following: he showed that *H. brasiliensis* is the source of the best quality rubber and he discovered the excision method of tapping to be the most effective way of extracting latex. He determined the best time to tap, and developed a method for the coagulation of latex. He also studied the agronomy of the rubber tree and its pests and diseases. Most importantly, he persuaded coffee planters to switch to rubber production. The first plantation was established in 1898 and Malaya quickly became the principal source of rubber for the world's industry, a position which it has never given up. Malaysian production is all based on the original Wickham stock of unselected heterogeneous wild genotypes (Purseglove, 1982b). Attempts to establish plantations in the centre of origin of the genus, in Amazonia, all failed, principally because of South American leaf blight, caused by the fungus *Dothidella ulei*. This fungus, endemic to the distribution range of wild rubber, causes little damage to the wild trees which are scattered at low density in the forest. Presumably they are shielded by the associated forest species which act as spore traps and protect the rubber from severe infections. When planted in uniform plantations, however, the rubber loses this supposed protection derived from dispersion and ruinous epidemics ensue. In consequence, Brazil produces only about 1 per cent of the world's rubber and even this is in an area distant from its origin. To the enormous benefit of Malaysia, and of the world's automobile industry, the seed material transported to southern Asia via Kew did not carry with it spores of *Dothidella*, and to this day Asian rubber has remained free of this crippling disease (although it is subject to other fungal leaf diseases endemic in the area).

In the early years of this century, workers investigated trunk-girth and productivity with a view to improving latex yield. Collection of seed from selected high-yielding trees was begun as a means of improving plantation stock but was abandoned when the feasibility of bud-grafting from proven high-yielding trees onto seedling root-stocks was demonstrated in Sumatra. This procedure was widely adopted in Indonesia, Ceylon and

Malaysia. Plantings came to be dominated by a few wild genotypes, the less productive wild trees were eliminated and the first clones roughly doubled the yield of rubber, In the 1920s and 1930s, breeding began with controlled pollination of selected parents and the seedling populations which it generated were used for direct planting and also as sources of second-generation clones. The best of these nearly doubled the yield again.

However, as might be expected, the loss of genetic diversity which followed inevitably from the selection for uniformly high yield, has brought its problems. Damage from leaf diseases is increasing and so is susceptibility to wind damage which is correlated with the increased yield (Wycherley, 1976).

In South America, breeding programmes for the incorporation of resistance to *Dothidella*, derived from wild species, are likely to be long and complex because of the difficulty of combining acceptable yield and quality of latex with high levels of resistance. Thus, rubber breeding in both its adopted and native habitats has rapidly reached the position which is common in other species, where breeders are turning to the genetic resources of wild populations and related species for genetic variability with which to improve the crop.

New crops are needed for different purposes

There is much interest at present in the development of new crops (Wickens *et al.*, 1989), and we can recognise various motives for this. There is the need for:

Alternative crops to replace those which are in overproduction, as is the case for most major crops of western Europe.

New crops to extend farming into hitherto marginal or hostile environments, in order to increase total food production.

The improvement of cultivated species which are scarcely different from wild plants. This is the case for a great number of species grown by poor subsistence farmers in tropical areas to provide nutritional supplements and dietary diversity. The potential here for 'new' crops is immense.

New crops to fulfil new needs. Rubber, sugar beet and jojoba fall into this category.

It seems scarcely necessary to reiterate the point that progress in any of these areas is dependent on the availability of genetic diversity. But our three examples of sugar beet, rubber and Triticale illustrate a point which is critical to the success of potential new cash crops. It is that the availability of genetic diversity is not in itself enough to ensure success. There needs to be a strong demand for the product. In the cases of sugar beet and rubber, there was an eager and growing market, whereas Triticale has had to look for a niche and is still struggling to find a place in cereal grain production which is commensurate with its potential. The fact is that attempts to produce new crops are to some extent moving against the general agricultural trend which is to narrow the crop base of the world's agriculture. However, the dependence on high-input systems is more and more being questioned and the need to develop sustainable systems based on new principles, techniques, varieties and crops may offer the incentives needed for the exploitation of a wider spectrum of the diversity of the plant kingdom.

4

Genetic diversity: what needs to be collected?

Résumé

Genetic resources of most field crops are usually collected as seed. The appearance of the seed, or of the mother plant at the time of seed collection, provides little information about the genes hidden within. Therefore collecting must be done 'blind' with regard to its purpose of assembling genetic diversity. Collectors should sample the seeds from a plant population in such a way as to maximise the chances of obtaining, in those seeds, a representative sample of the genetic diversity of the population.

We can recognise three levels of genetic diversity: differences between alleles of a gene, differences between the genotypes of individuals within a population and differences in the frequencies of alleles between different populations. Which should we aim to collect? Theoretical considerations lead to the conclusion that we should collect common alleles and adapted genotypes from as many different populations as can be identified, collecting the required number of seeds – 4000 has been recommended – from fifty to one hundred plants per population, and with approximately equal numbers of seeds from each plant.

These standards are relatively easy to achieve in most cultivated species because of the large numbers of plants which are usually grown in a crop, but difficult in garden and back-yard cultures where numbers are small, and in many wild species where population densities vary from species to species and, in annuals, from year to year. For species distributed over large geographical areas, it is often difficult to define the boundaries of individual populations and, for many species groups, there are frequently problems in the identification and naming of the constituent species.

Priority in the past has been given to the collection of landraces and old varieties and for the major staple crops. Much of the surviving material in these categories is thought to be now in store, and attention is turning to the collecting of wild species which are related to the major crops.

There are large numbers of samples of cultivated forms in existing collections. Only rough estimates are available of numbers of unique accessions and, with some notable exceptions, there is often little information on the characteristics of this material or on its viability. Extensive analysis is needed, crop by crop, before we can say whether more collecting is necessary.

It is a fundamental principle observed by most of the genetic resources community, that germplasm and its associated data should be available without restriction to those who use it for the benefit of mankind. Given the operation of this principle, it is not then necessary for each germplasm collection to be comprehensive. Each can be regarded as a component part of a dispersed world collection which can become a functional reality through the formation and operation of collaborative world crop networks. Joint analysis of the accessions of all the collections in a network will reveal the true extent and completeness of the world collection and the need for further collecting. Crop germplasm networks are currently being established for a number of major and minor crops.

For some species it is necessary to collect vegetative material in the form of shoots, buds or embryos. There are three cases where this method is necessary: when the crop or species is sexually sterile, when it is convenient to breeders to do so (as in the case of many tree fruits) and when seeds can be stored only for very short periods before death ensues, because of intolerance to dessication or low temperatures, or both (recalcitrant seeds). For the collection and storage of materials of these types, *in-vitro* culture methods are being developed and their effects on the germplasm are being studied.

Seeds are ideal natural structures to collect and conserve

Most crop species produce seeds and from the point of view of collecting and conserving genetic resources, seeds are highly convenient units in that they have a full set of genetic information inside an embryo, with associated food reserves and the whole usually occupying a small volume. There are some exceptions to this general rule. For example, the seeds of Faba and *Phaseolus* beans, and of many forest tree species, are capable of storage at low temperatures but are inconveniently bulky. Against this natural convenience of seeds for collecting and storage one has to set the disadvantage that they provide few clues from their appearance to the genes which they contain. The appearance of the seed-bearing plant usually permits its identification at the species level, but otherwise it provides limited information on its genetic constitution and that of its seed progeny. So there is a problem; the collecting of seeds has to be done without knowledge of the genetic material which they contain and yet their genetic diversity is the reason for the collecting and subsequent use.

Genetic diversity has different levels of organisation

Differences in the overall appearance (morphology) and functions (physiology) of plants are frequently attributed to differences in genes. While this may be largely true if we are comparing plants belonging to different species and genera, for plants within a species it is misleading to use the term 'gene' in this context. Two plants from the same species, which may be clearly different in their appearance, for example, in height or flowering date or in seed or flower colour, probably have the same gene complements. The matter centres on what we understand by the term 'gene'.

A gene is a functional sub-unit of the plant's DNA which has a coded set of information for the production of a particular protein, which may be a structural protein, an enzyme or a storage product. Corresponding sub-units in the DNA of two different plants of the same species will code for the same protein. That is to say, both plants will have the same gene in the same place. However, genes can exist in slightly different alternative states called alleles and most if not all genes have many different alleles. The alleles can differ in the quality or the quantity of protein which they produce, or indeed in whether they produce any or none at all.

Most differences among plants within populations are due to the many possible combinations of the alleles of all the genes in the total genetic complement (usually referred to as the genotype).

Differences between populations of the same species are due to differences in the frequencies of the different alleles of the genes. Alleles can have a high frequency in one environment and low in another as a result of natural or artificial selection, acting either for or against the allele according to its effect on the adaptive fitness of the plant. Adjustment of allelic frequencies through selection is the basis of the adaptation of a population to its environment.

We have identified three levels of genetic diversity: between the alleles of a single gene, between combinations of alleles of different genes in the genotype (differences between individuals within populations), and between populations in different habitats (differences between adapted allelic complexes).

Which types of genetic diversity should we collect?

We cannot conserve all genetic diversity of crops, and therefore all collecting activities involve the sampling of what exists in the field or in the wild. Sampling raises problems of choice between populations and between plants within a population if the population is heterogeneous, as is the case in wild species and outbreeding crops. Collectors bear a heavy responsibility, because their decisions on what and how to collect determine what will be available to posterity. Considerable attention has been given to the problem of how to collect most efficiently within populations. We should consider some of the arguments which have raged on the two central issues; what should we be collecting and how should we do it?

Should we collect 'useful' alleles?

In the light of the considerations above, on the nature of genetic differences between individuals, the first of the two central issues can be restated as: what should we collect in terms of alleles? There have been two points of view on this matter. The first says that we should concentrate on collecting 'useful' alleles. Unfortunately this apparently sensible aim does not stand up to more than a moment's critical examination. For example, it is scarcely possible to define what might be of use to a breeder because the needs of current breeding programmes are so diverse and the needs of the future are impossible to predict. Two insuperable difficulties settle the issue. In most crops information is lacking on the numbers of alleles which exist for each of the genes of the genome. Even where this information does exist, the presence or absence of a particular allele can rarely be deduced from an inspection of the plant's appearance (phenotype), at the time of the seed collection. Also, differences between individuals in a population can often be due to environmental factors – local differences in soil fertility or soil moisture, for example, and therefore phenotype can be a very unreliable guide to genotype.

Should we collect rare or common alleles?

The second contentious topic concerns whether we should aim to collect common or rare alleles. Even in an ideal world it is doubtful whether we should aim to collect a copy of each allele which exists in a population. To do this with a reasonable probability of success, bearing in mind that the collecting must be done 'blind' – without knowledge of how many alleles there are or where they are – would involve the collection of inordinately large numbers of seeds from each population. Certainly, far more than could be stored and managed in genebanks! Some scientists have worried that rare alleles are likely to be missed by the 'blind' sampling and therefore lost to posterity. There appears to be an unstated assumption that rarity is equated with high value, a basic principle of the antiques business, but without foundation in relation to plant genetic diversity. Rare alleles in a population are rare because they are disadvantageous in conferring lack of fitness to the environment in which the population lives. The same alleles may be advantageous, however, in another environment where, occurring at a higher frequency, they may be collected more readily. Thus, in order to fully sample the allelic diversity of a species it is very important to collect from all the different environments in which a species occurs across its distribution range. This point is illustrated by the collection of alleles for resistance to grassy stunt virus in rice (discussed in Chapter 3), which were found in only one of many thousands of accessions tested, but in this one they occurred in a moderately high frequency.

Should we collect alleles or genotypes?

A more realistic dispute has centred on the problem of whether we should aim to collect alleles as single genetic entities or whether we should aim to collect adapted assemblies of alleles (genotypes). It is true that a significant number of plant characteristics are determined by single alleles, for example pigmentation variants of flowers and seed and, in particular, some, but not all, disease and pest resistances. However, many plant characters are determined by two or more genes each varying in its contribution according to which of its alleles are present. Finally, the whole plant can be regarded as the ultimate expression of all the interacting systems of genes, which in turn are subject in their expression to the influence of the environment. The influence of any one allele, therefore, on the appearance or fitness of the plant is usually small and the fitness of an individual can be more realistically attributed to the function of its whole genotype.

In their search for superior parents, plant breeders are not only interested in particular characters, such as disease resistances or grain colour, but also in the selection of parents with general adaptation to the type of environment for which their new varieties are intended. These points lead to the conclusion that we should aim to collect a representative sample of the adapted gene-complexes from populations in distinct ecological or agro-ecological habitats. In so doing, we shall simultaneously sample alleles which are present in significant frequencies in the different habitats and thus assemble a representative collection of the genetic diversity of the species.

Theoretical studies provide a rational basis for effective collecting

How then should the collecting be done? What principles should we adopt in sampling genetic diversity? These questions have received much attention from numerous workers and notably from Marshall and Brown (1975). With a combination of population-genetics theory and practical realism, they proposed that collecting should aim to capture at least one copy of each common allele; a common allele being defined arbitrarily as one with a frequency of more than 5 per cent in the population. Since data on allelic frequencies in populations are usually lacking, they suggested, on the basis of their calculations, that the best way to achieve this objective would be to collect equal numbers of seeds from fifty to one-hundred individuals in each population, and to collect from as many diverse sites as possible so as to maximise the possibility of collecting genotypes with different adaptations. The minimum plant numbers required per population or site, to capture most of the genetic diversity, appear therefore to be quite modest. The Marshall and Brown recommendations provide a valuable reference point for collectors who are forced by circumstances in the field to change their plans or modify their collecting procedures.

Collectors have other factors to consider when deciding on the number of seeds to collect. The sample size has to be adequate to meet all the immediate needs, particularly for viability testing to check the proportion of the seed which is alive, and for distribution of duplicate samples to different genebanks. Seed will be needed to produce plants for characterisation and other descriptive studies and, if the collecting has been done by expatriates, half of the collected material should be deposited in a genebank in the country in which it was collected. Consideration of these factors led the IBPGR to recommend a minimum sample size of 4000 seeds. When collecting seeds of landraces or old cultivated varieties, either by gathering from growing crops or by purchase from rural markets, the recommended sample size is easily obtained. With small populations and especially for wild species this number can often be unattainable, in which case seed multiplication has to precede seed storage.

Wild species present special problems

The problems inherent in collecting wild species arise from the variability of wild populations. The differences in population characteristics of different species and the population variability from year to year prevent the formulation of simple general rules for collectors. For example, the density of plants may vary from hundreds per square metre, as in the case of some annual grasses, to thousands of square metres per plant, as in the case of temperate and tropical tree fruits. Population densities of annual species can vary between wide extremes from year to year according to factors such as rainfall or grazing pressures, with consequent effects on seed numbers available for collection. A common feature of wild species is their possession of an effective seed dispersal mechanism, and in many annuals seeds are shed and distributed, sometimes over large areas, in a short period of time. The timing of arrival of the collector can therefore be quite critical; too early, and the seeds are immature, too late, and the collector is reduced to searching for seeds on the soil surface or down fissures. Either way, the collection of an adequate random sample is impossible. Sometimes, an unsuccessful first visit can be used to predict the timing of more productive fieldwork in the following year, but interesting wild species often grow in places which are either geographically remote or inaccessible for political reasons, or both, so that return visits to unproductive sites can be difficult to organise.

There are often real difficulties in deciding on the spatial limits of a population and hence on the limits of a sampling area. With small, discrete localised populations, the limits will be obvious, but in many cases considerable experience is required to synthesise observations on habitat – on soil type, slope, aspect, altitude and associated species – into a sensible subjective decision on what constitutes a population. The Marshall and Brown principles, briefly presented above, are a valuable theoretical framework against which practical collecting tactics and the inevitable enforced compromises can be judged.

The description and classification of plants presents tricky problems in many genera, as for example in the presumptive wild grass ancestors of cereals such as wheat, oats and rice, where large differences in genetic composition can be concealed behind external forms which are superficially very similar. The science of taxonomy aims to bring order to the confusion of natural diversity, by grouping together those plant types which show similarities to each other and particularly in the morphology of their reproductive organs. This has led, in many cases, to the recognition of a plethora of taxonomic categories – species, sub-species and others – often separated only by minor differences in phenotype. Furthermore, the taxonomic schemes of different authorities not infrequently conflict with each other. The problem therefore of confidently identifying and naming material in the field can be formidable for a collector who is not a specialist in the group and sometimes for those who are.

From the point of view of genetic resources conservation, the emphasis naturally needs to be placed on genetic relationships rather than on similarities of appearance. Taxonomic schemes with this genetic bias aim to recognise 'natural' or 'biological' species, which in broad terms are separated by barriers of inter-sterility; gene exchange through sexual reproduction is possible between entities within a species, (hybrids are fertile), but not between species (hybrids are sterile or inviable). One highly desirable consequence of this genetic approach to taxonomy is that it often reduces the numbers of species which have to be considered. A major limitation is that the necessary information on crossability of taxa is usually lacking.

The collecting of seed of wild species is much more complex than the relatively straightforward collecting of landraces and primitive varieties. The collecting of seed samples of wild species is best carried out as part of a larger research project concerned also with related topics such as the evolutionary pathways of species within a genus and their eco-geographic distributions. Many crop species groups are poorly understood, especially so in the case of tropical crops and acutely so in the case of tree crops, and much research remains to be done on them before their genetic resources can be adequately collected. This research in field and laboratory usually proceeds in a stepwise manner. Studies on the first collections lead to further, more discriminating field forays until a complete picture emerges of the diversity in a species group and of the eco-geographic distributions of its constituent taxa.

To sum up; with regard to the wild relatives of crops, in many cases we do not yet know with certainty what many of them look like, or what needs to be collected nor where we should go to find them. And yet, because of large-scale destruction of habitats which we know to be taking place, some, perhaps many, species must be under serious threat of erosion if not outright destruction. In these cases, conservation cannot await the results of research. Urgent rescue collecting is needed. Pragmatic decisions based on a rapid synthesis of available, if incomplete, knowledge may be required as a guide to rapid action, supplementing rather than replacing the long-term scientific studies.

Priorities in seed collecting

In the decade following its foundation in 1974, IBPGR, in collaboration with FAO and with scores of institutes and hundreds of scientists throughout the world, was directly and indirectly responsible for the collection of large numbers of seed samples and for their deposition into national seed collections and into those of the International Agricultural Research Centers. This first phase of collecting was motivated by the urgent need to rescue landraces and old varieties before they were lost for ever and to conserve them in well-managed collections.

IBPGR recognised from the outset the need for a collecting strategy to guide its allocation of funds and technical support to areas of greatest need. Its decision to concentrate on the major staple crops reflected the world food shortage and the wishes of its donors. These crops were classified by region and country into first, second and third priorities. Priorities were assigned according to known or predicted probability of serious erosion of diversity. As work progressed during the first decade of operations, priorities were revised in the light of progress made and of newly identified needs.

It seems likely that, of the landraces and primitive varieties which had survived to this period, a significant proportion were rescued, but this can be no more than an inspired guess because there were no detailed data against which success could be measured.

Cultivated forms rightly make up the greater part of the collected material since their rate of loss was clearly greater than that of wild species. Now, with much of the surviving primitive cultivated material rescued, attention has turned to the selective filling of gaps in collections and to the collection of wild and weedy species.

Present status of collections

Numbers are a poor guide to progress

In terms of numbers of accessions taken into collections, the first fifteen years of internationally guided collecting appear to have been spectacularly successful. The most recent data (Table 4.1), show that there are more than three million accessions in national collections and in those of the International Agricultural Research Centers. More than half of these are accessions of the staple cereals.

It is known that replication of the same material in different collections is common. Partial data for a few crops indicate that this may be as high as 40 per cent but the true extent is unknown and therefore the number of unique accessions for each crop has yet to be determined.

The large numbers of accessions of crop germplasm which are now conserved in genebanks is impressive and superficially encouraging. It

Table 4.1. *Total numbers of collections and samples of germplasm of major crops and crop groups. (Extracted from IBPGR database, 1990)*

Crops	No. of collections	Total no. of accessions
Cereals		
Barley	63	256 651
Buckwheat	3	5 247
Millets	25	33 101
Maize	61	152 010
Oats	37	109 117
Rice	45	343 296
Rye	14	12 707
Sorghum	34	137 799
Teff	3	4 312
Wheat	115	508 892
Total	400	1 563 132
Grain legumes		
Beans – *Phaseolus*	65	158 098
Vicia	38	37 836
Chick pea	26	47 670
Groundnut	40	66 074
Lentil	25	19 801
Mung bean	34	69 714
Pea	33	52 015
Pigeon pea	9	19 949
Soybean	70	136 680
Total	340	607 837
Roots and tubers		
Aroids (Taro, Eddoe and Tannia)	39	6 474
Beets	37	13 984
Cassava	41	25 071
Potato	90	62 909
Sweet potato	56	25 714
Yams	31	9 799
Total	294	143 951
Vegetables		
Aubergine	7	2 924
Brassicas	43	40 351
Carrot	5	3 751
Celery	1	250
Cucurbits[a]	58	46 095
Lettuce	10	8 314
Okra	7	5 470
Onions, leeks, etc.	17	11 252
Peppers	39	36 411
Radish	8	4 677
Spinach	4	1 252
Tomato	39	57 735
Winged bean	9	4 740
Total	247	223 222

Table 4.1. (*cont.*)

Crops	No. of collections	Total no. of accessions
Fruits – tropical		
Tropical fruits	36	9 599
Tropical nuts	10	3 729
Avocado	16	2 941
Banana	31	5 408
Citrus	52	16 992
Durian	5	1 160
Mango	28	4 704
Passion fruit	1	194
Pineapple	7	774
Soursop	4	383
Total	190	45 884
Fruits – temperate		
Almond	9	1 169
Apple	62	50 562
Apricot	17	2 641
Blueberries	3	1 393
Cherry	25	10 269
Chinese gooseberry	1	105
Currants	9	1 653
Fig	7	1 871
Mulberry	2	289
Peach	40	8 212
Pear	41	9 427
Persimmon	4	663
Plum	23	3 183
Prunus	15	5 376
Quince	3	240
Raspberry	9	1 530
Strawberry	14	3 788
Temperate nuts	20	5 678
Total	304	108 049
Fibre crops		
Cotton	37	43 180
Other fibres[b]	11	15 589
Total	48	58 769
Oil crops		
Oil crops	8	9 680
Oils – edible[c]	15	37 032
Oils – industrial[d]	11	9 678
Total	34	56 390
Forages – browse		
Browse plants	2	675
Atriplex	1	900
Leucaena	4	2 080
Total	7	3 655

Table 4.1. (*cont.*)

Crops	No. of collections	Total no. of accessions
Forage grasses		
Forage grasses	37	76 871
Agropyron	4	3 470
Bromus	5	4 080
Dactylis	11	8 715
Elymus	2	1 451
Festuca	9	7 074
Lolium	11	9 505
Panicum	14	16 843
Stipa	1	251
Total	94	128 260
Forages legumes		
Forage legumes	17	25 059
Astragalus	1	285
Centrosema	3	3 421
Desmodium	3	3 466
Trigonella	2	606
Lotus	4	3 054
Medicago	20	36 163
Onobrychis	5	2 004
Stylosanthes	5	6 958
Trifolium	25	34 395
Total	85	115 411
Summary		
Cereals	400	1 563 132
Grain legumes	340	607 837
Roots and tubers	294	143 951
Vegetables	247	223 222
Fruits – tropical	190	45 884
Fruits – temperate	304	108 049
Fibre crops	48	58 769
Oil crops	34	56 390
Forages – browse	7	3 655
Forages – grasses	94	128 260
Forages – legumes	85	115 411
Total	2043	3 054 560

Note: Cereals, vegetables, grain legumes, forage grasses, forage legumes, cotton and oil crops include collections with 200 or more accessions. Other groups include collections with 50 or more accessions, except for tropical nuts, where the cut-off points is 20 accessions.
[a] Cucurbits include watermelon, melon, cucumber, pumpkin, squash, marrow.
[b] Other fibres include hemp, flax and sisal.
[c] Oils – edible include palm, olive, safflower and sunflower.
[d] Oils – industrial includes castor oil.

might lead to the conclusion that the genetic resources of our major crops are secure and that the crop genetic resources component of the wider conservation movement has been remarkably effective and has largely completed its work. It is true that it has done a great deal, but only part of what needs to be done. We recall that the purpose of collecting and conserving crop diversity is to provide genetic resources for use by plant breeders, and simply putting seeds in a cold store or plants in a Field genebank is not sufficient to make the material available for use.

The simple numerical data, on numbers of accessions in store (Table 4.1), is valuable in providing a general impression of the magnitude of germplasm already conserved but is of limited practical use. Potential users need to know where to go in the world to obtain germplasm suitable for their purposes. For this they can turn to the series of comprehensive *Directories of Germplasm Collections* published by IBPGR. These list, by crop and country, the names and addresses of curators and information on their collections including, ideally, the numbers of accessions of cultivated and wild species, their countries of origin and information on the storage conditions, documentation, quarantine restrictions and availability of seed. A specimen page from the *Directory of Germplasm Collections 4. Vegetables* (Bettencourt and Konopka, 1990), is reproduced as Fig. 4.1. This particular entry in the *Directory* is admirably comprehensive, but it is common for one or more of the categories of data to be unavailable to the compilers of the *Directory* and therefore entries are all too often incomplete.

The *Directories* provide a valuable guide to where to look for material, but are insufficient to enable potential users to specify which particular accessions are most suited to their purposes. Here, they must define their needs to the most appropriate curator, as determined from the information in the *Directory*, and ask for seed of what appear to be the potentially most useful accessions, selected by the curator from the more detailed information on the biological characteristics of the accessions which he or she has in the database.

It is a sad but perhaps unsurprising fact, that as the data we need become more detailed, they also become scarcer. Gaps in *Directories* are exceeded by the gaps in genebank databases. Some crop collections are well documented but many are not. All too frequently, data on viability of accessions, on their geographical and ecological origin, and on their genetic, agricultural and disease-resistance characteristics are lacking. Most frequently, the degree of documentation varies between different parts of the same collection. But the general consequence is that, the fewer

PHILIPPINES

National Plant Genetic Resources Laboratory (NPGRL)	Telephone: (63) 833 35 28
Institute of Plant Breeding (IPB)	Telex: 3432 PTLBK PU
University of the Philippines at Los Baños	Cable: PLANTBREED PHILIPPINES
College, Laguna	

Curator/person in charge: N.C. Altoveros

Details of collection:
Lycopersicon cheesmanii, 2 wild from Ecuador and Galapagos
L. hirsutum, 9 accessions from Ecuador (5) and Peru (4)
L. lycopersicum, 4306 accessions from Argentina (54), Bolivia (76), Brazil (97), Canada (125), PR China (85), Colombia (81), Costa Rica (55), Czechoslovakia (67), Ecuador (91), El Salvador (399), Guatemala (201), Honduras (82), Hungary (163), India (99), Iran (60), Italy (57), Mexico (80), Peru (180), Philippines (220), Taiwan (China) (99), Turkey (191), USA (839), USSR (75), Yugoslavia (113) and 717 from 53 other countries
L. lycopersicum var. *cerasiforme,* 28 accessions from 10 countries
L. peruvianum var. *peruvianum,* 67 accessions from Chile (17), Ecuador (3), Peru (44), Poland (1), USA (1) and 1 of unknown origin
L. pimpinellifolium, 182 accessions from Ecuador (27), Mexico (22), Peru (111), USA (16) and 6 from 6 other countries
Lycopersicon spp., 12 accessions from USA (9) and 3 of unknown origin
L. hirsutum x *L. glabrosum,* 4 accessions from Ecuador and Peru
L. lycopersicum x *L. hirsutum,* 2 accessions from the Netherlands
L. lycopersicum x *L. peruvianum,* 3 accessions from USA
L. lycopersicum x *L. pimpinellifolium,* 176 accessions from 31 countries
L. peruvianum x *L. hirsutum,* 2 accessions from Peru

Maintenance of collection:
Long-term seed storage at -20ºC
Medium-term seed storage at 5ºC and 40% RH
Seeds with 6% m.c. in glass bottles, tin cans or in aluminium laminated foil packets

The Institute has agreed to accept responsibility for maintaining an Asian collection of *Lycopersicon* and a Southeast Asian vegetables collection for long-term conservation

Duplication of collection: Partially at AVRDC, Taiwan (China) and USDA, USA

Availability: Available on exchange basis

Quarantine: Import permit required

Evaluation: On-going. Around 500 accessions are evaluated every year

Documentation: Complete and computerized for passport data. On-going computerization for evaluation and characterization data

Fig. 4.1 Specimen of crop germplasm directory information. (Reproduced from Bettencourt and Konopka, 1990.)

the data, the less chance an accession has of being selected by a breeder for use as a parent in a breeding programme.

The frequent lack of data is regrettable but explicable. Early collections came into being in different ways and for different purposes, and in general the extent of documentation was no more than appeared to be necessary to meet the needs of the immediate user. Furthermore, it was common for seed to be exchanged without the data which related to it. The systematic assembly of comprehensive data in databases is a relatively new development.

Some collections were set up by assembling material from breeders' working collections. Breeders had traditionally exchanged materials among themselves and the numbers of duplicates were usually large but unrecorded. Others were collections made to provide material for research programmes on, for example, species relationships and evolutionary pathways. Yet others, while principally for breeding or research, contained a significant number of samples specifically collected for genetic resources conservation. Notable among these are the Vavilov collection in Leningrad and others in Europe, the collections of some staple crops assembled by the Rockefeller Foundation and the US National Science Foundation and those collections related to colonial development, such as those of cotton and banana funded by the United Kingdom.

Further confusion arose because some genetic resources programmes and collections grew out of long-standing plant-introduction activities where the principal purpose was to find new crops and better varieties of old ones. It was common for unsuitable material to be discarded but, even if retained, the documentation was usually minimal.

The frequent lack of documentation is therefore understandable but none the less regrettable since it is often the old and highly variable landraces which are no longer available from the field which are most poorly described.

Consideration of these large numbers of samples leads immediately to the question, do we already have enough germplasm conserved? Does it adequately represent the diversity, which has evolved over millenia and which is under threat in the field or in the wild, or do we need to collect more? The only possible way in which we can arrive at a satisfactory answer to these important questions is to make a qualitative analysis of the data attached to the material which is being conserved – to find out what it is, rather than simply to count how much there is of it, and to relate this to what we know, or to what we can guess, about diversity which is under threat of erosion or extinction. Work has begun along these lines on some crops in some collections, but it is largely uncoordinated except for new developments within crop networks, discussed below (see pages 126–128).

Qualitative analysis of collections is a most urgent task

The key to understanding the nature of material in collections lies in the initial data relating to each accession, the so-called Passport data. These should include accession identifiers given by curators and information recorded by collectors, such as variety and species names, donor name when the material was received from another institution or individual, acquisition date, numbers of seeds, collector's number, date of collection, country and exact location of collection site including longitude, latitude and altitude and, in the case of wild species, information on the habitat in which the collection was made. Much material, particularly that which was collected in the early days, has little data attached to it and, in the case of some landraces and primitive varieties, nothing more than a name.

The analysis of whatever data there are, provides the only possibility of understanding what we already have in collections and of formulating policies for future action. As a first step it permits the determination for each collection of the identity of the unique samples of each crop or species and hence of the amount of replication within the collection. The analysis may also reveal the degree to which the collected germplasm represents the known distributional range of the crop or species. By collaborative action between curators it would be possible to widen this understanding to obtain a world-wide overview of the nature and replication of conserved diversity.

When the analyses are as complete as the limitations of the data permit and when we understand the extent to which the collected material represents the range available in the wild, the basic questions remain; how large does an individual collection need to be, do we need to collect more than we already have and should each collection aim to hold a comprehensive representation of the variability which exists in the crop or in the wild?

The costs of collecting are relatively small compared to the subsequent activities in the genebank, such as the storage, seed viability monitoring, characterisation and seed multiplication of the accessions, which make permanent heavy demands on resources. Some national funding authorities may decide for reasons of convenience or status to establish a fully comprehensive collection and to accept the costs involved, although, not infrequently, adequate financial provision is overlooked, to the detriment of the genebank's function. However, from the point of view of germplasm conservation, it is not necessary for all collections to be comprehensive and it can be persuasively argued that it would be a serious waste of resources if they were.

The alternative is to develop a well-defined integrated network of collections for each crop, where participants share work, responsibility, information and of course the genetic resources which they are collectively conserving.

Further collecting requires collaborative action

The guiding principle that crop genetic resources should be freely available to all with a serious interest in their use has been widely observed by scientists and plant breeders in the past. Nevertheless, it is necessary to emphasise it again because it is essential to the functioning of germplasm networks. Recently, some governments have adopted, as a national policy, either restrictions on, or outright banning of, export of genetic resources from within their jurisdiction. Fortunately, these cases are relatively rare and do not accord with the international consensus.

Given the operation of the unrestricted exchange principle, it is possible to regard national collections as components of a dispersed world collection of a crop's genetic diversity, and from the acceptance of this concept, many practical benefits can ensue. Participants in a crop genetic resources network could work together to determine the numbers and identities of unique accessions and the degree of redundancy among collections. The contents of collections could then be rationalised by the relegation of unwanted redundant accessions to an inactive reserve collection if so desired, and the size of active collections thereby reduced, in order to utilise resources more effectively. Joint analysis of accession data to determine the extent to which they represent the eco-geographic range of the crop would permit an assessment of the need for further collecting to fill gaps in the world collection.

If we now repeat our original question – what more needs to be collected? – we can see that a rational answer can only come when the crop networks have completed the analytical work on the material already in collections. The alternative, which does not bear contemplation, is for each collection or national programme to work in isolation, to base its independent operational policy solely on national considerations and to assemble a fully representative collection of the genetic diversity of each crop gene-pool. The potential in such a policy for waste of financial and human resources is enormous and it is an approach which has been partly responsible for the present unsatisfactory state of plant genetic resources conservation. The issue of crop networks is discussed more fully in Chapter 6.

The collecting of vegetative material is sometimes necessary

Seeds are such convenient units of collection and storage – generally small in size, of low moisture content, damage and rot resistant, dormant and usually abundant – that collectors only turn to collecting vegetative material under special circumstances. Sexual sterility, as in banana varieties, is of course the most powerful reason – in the absence of seeds there is no option but to turn to vegetative parts which can be used for propagation, such as cuttings of stem and root, tubers, buds etc. This is the case with much genetic resource material of local varieties of yams and sweet potato. In other species, such as coconut and avocado, the seeds are too large for their convenient collection. The same difficulty can arise with the collection of vegetative parts, for example the tubers of yams commonly weigh between 5 and 10 kilograms. Also in some fertile species with seeds of a covenient size, it may be impossible to store the seed for more than a short period; the seeds are said to be recalcitrant. Examples are rubber, cocoa, many tropical tree fruits, and some tropical and temperate timber species (Roberts *et al.*, 1984). There have been attempts to devise simple practicable methods for improving the storage life of recalcitrant seeds, but so far there have been no major breakthroughs. The cases quoted above can be regarded as examples where enforced vegetative collecting is necessary.

There are other circumstances when the collector may choose to collect vegetative material in order to preserve the exact combinations of alleles which determine particular clonal varieties of say potatoes or tree fruits. Clonal conservation is sometimes desired because sexual reproduction disrupts the combinations of alleles in the parents, so that progeny always differ from them and from each other – except in strictly self-fertilising species which 'breed true'. It is undoubtedly true that the genetic resources of some tree fruits such as apple and pear may be conserved economically as seed, but there is a considerable penalty to pay when the time comes to use them. Seedlings must be grown for several years before they emerge from their juvenile phase, their fruit characters can be assessed and judgements made on the potential value of the individual seedling as a parent in a breeding programme. Accessions maintained as mature trees in Field genebanks can be fully documented and can provide pollen to breeders every year. On the other hand, and in common with most vegetative collections, they suffer from the severe disadvantages of small sample size and liability to accumulate infection by disease organisms.

Collectors not infrequently turn to collecting vegetative material of perennial species when there is no seed available at the time of their visit. Judging correctly the time of arrival to coincide with seed ripeness is one of the more difficult aspects of collecting and the ability to collect vegetative specimens – buds, shoots, tubers or roots, which can be used for propagation, can obviate the need for a second visit.

The disadvantages of collecting vegetative material are small sample sizes and disease. Living material removed from the plant is usually subject to fungal and bacterial decay, particularly in warm, humid climates. It needs to be taken with the minimum of delay to a nursery or glasshouse where it can be rooted or grafted under conditions favourable to growth. There is the other important practical point that vegetative material is subject to stringent quarantine controls in most countries. These necessary controls are liable to hold up the passage of material, in some cases for many years, when the quantity of material exceeds the capacity of the quarantine system.

Recent developments in the collecting of material directly into tissue culture offers possibilities for circumventing some of the problems outlined above. For example, a method has been field tested in Cote d'Ivoire and Indonesia for the collection of coconut embryos. The embryo is removed from the nut by the collector in the field, placed directly into culture in a small glass tube, and the bulky nut is discarded. Success rates are high in terms of the numbers of established seedling trees obtained (IBPGR, 1989). Other projects are concerned with developing and field-testing methods for the collection of woody material of stone fruits, grape vine, cocoa, avocado, citrus and breadfruit and of vegetative parts of cassava and of tropical and temperate forage grasses (which often lack seed heads due to grazing) (Withers, 1989). In all cases the aim is to facilitate collection and to lengthen the life of collected material so as to increase the probability of its arrival in a viable state at the genebank. The culture medium includes not only nutrients to support the plant parts but also contains fungicides and antibiotics to suppress micro-organisms which cause deterioration. Procedures are simple and equipment is minimal but the effectiveness of collecting vegetative material is greatly improved.

The collecting and maintenance of genetic resources as vegetative material in field genebanks is expensive and subject to many disease hazards. It is usually practised when no other method is available or, alternatively, when the costs are thought to be justified by the convenience to the breeder, as in the case of many tree fruits.

Undoubtedly the preferred method will continue to be the collecting of seed and, as the emphasis moves progressively from cultivated to wild species, it is likely to become more and more a task for scientific specialists with expert knowledge of the species group.

5

Safeguarding genetic diversity

Résumé

The safeguarding of genetic diversity after it has been collected is a matter of great importance; genetic erosion is a constant threat inside genebanks as well as outside them. Conserved genetic diversity is intended to meet the needs of breeders now and in the future. Current use requires accessibility, posterity requires long-term security, and the two are difficult to reconcile in the same batch of seed. Accordingly, seed samples are divided into two: one part stored at around 0 °C forms the Active collection for immediate use in normal genebank activities and the other is stored at lower temperatures (preferably at -10 to -18 °C) and at 5 per cent seed moisture content and forms the Base collection for long-term storage.

Large seed stores operating at -10 to -18 °C in countries with high ambient temperatures are expensive to run and two possibilities for more cost-effective storage are under investigation: the location of stores in permafrost, where temperatures are remarkably stable at about -4 °C, and the pre-storage drying of seed to ultra-low moisture contents as a substitute for deep-freeze storage temperatures.

While seeds are particularly suitable for collecting and storage, not all species can be handled in this way. Some are sexually sterile and others possess seeds which cannot withstand drying, or low temperatures – or both – and are said to be recalcitrant. Occasionally it is thought desirable to maintain by vegetative means, particular clones of sexually fertile species, in order to preserve particular gene combinations. For these categories of material, conservation is usually as growing plants in Field genebanks. These are relatively costly to maintain, can require large areas of land and tend to accumulate infections by various disease organisms.

Possible alternatives under investigation are the storage of buds or tissue pieces under 'slow-growth' conditions in *In-vitro* Active Genebanks, (IVAGs) and the storage of excised pieces of tissue or embryos in liquid nitrogen at $-197\,°C$. The latter, if feasible, would constitute *In-vitro* Base Genebanks (IVBGs). Both methods offer protection from disease infections but the former may pose threats to genetic stability of the stored material. Neither has yet been fully worked out.

Ex-situ conservation has been widely criticised for supposedly preventing the normal process of evolution and because of the serious risks it introduces to the genetic integrity of the accessions during the process of seed multiplication. Both points are examined and assessed.

Conservation *in situ* has been widely advocated for crop genetic resources. The different requirements and procedures for crop plants and for their wild relatives are outlined. *In-situ* and *ex-situ* methods of conservation are complementary; *in situ* is appropriate to the prevention of erosion of wild species and *ex situ* to the conservation of species, cultivated and wild, which are already being destroyed. For both categories of material a monitoring and early-warning system is necessary, in order that timely action may be taken.

Collected germplasm has two purposes

Seed material which has been collected and brought into genebanks has two purposes: to provide genetic diversity for use by plant breeders of today, and to meet the needs of the breeders of the future. Collected material must be accessible for present use and secure in the long term. To a considerable degree these two requirements are in conflict if imposed on the same lot of material. The solution adopted has been to divide each accession into two parts and to manage each part differently.

Seed lots intended for long-term, secure storage are placed in Base collections, physically separated from the corresponding material in Active collections and managed so as to maximise the time interval between regeneration cycles. Regeneration is achieved by producing fresh seed of high viability and substituting it for the old seed in the Base collection, and the process of seed multiplication presents serious hazards to the genetic integrity of the accession, a matter discussed more fully in Chapter 6.

Seed in Active collections is used for the everyday activities of the genebank, which include documentation, characterisation, viability testing, seed multiplication and distribution for research and plant breeding purposes. Base collection seed would be used to replace seed in the Active collection which had been inadvertently lost or had its integrity compromised either by accidental mixing of seed, or by genetic contamination from foreign pollen due to inadequate isolation during growing-out for seed multiplication. If Active collections are functioning properly, it should rarely be necessary for the seed in the Base collections to be disturbed. A final security measure is the duplication of Base collections in other genebanks as an insurance against natural disasters or civil disturbances. The two types of collection have complementary and connected activities.

Seeds are naturally adapted for storage

Most crop genetic resources material which has been collected to date, is from field crops, vegetables and forages, and the greater part is stored as seed. Two factors which control the length of time which seed may be stored are its moisture content and the temperature of the storage environment. Harrington (1970) reports that below 14 per cent moisture content, successive reductions of 1 per cent water content double the life of the seed. Similarly between ambient temperatures of 50 °C and 0 °C every reduction of 5 °C doubles the life of the seed. These general rules have been refined more recently after detailed research on different species.

Although procedures can vary from one species to another, in general, if seed is dried slowly to a low moisture content (usually to 5–8 per cent), and subsequently sealed in moisture-proof containers, it is believed that viabilities can be maintained for between thirty and a hundred years, according to species, at − 18 °C, and up to thirty years when stored at 0 °C. The lower temperature is recommended for long-term storage of Base collections and for convenience and economy, Active collections usually operate at 0–5 °C.

Seed physiology research has provided clear guidelines on the preparation and storage of seed to ensure long-term viability. Information on relevant aspects of seed physiology is available for curators in publications of the IBPGR, for example the IBPGR *Handbook of Seed Technology for Genebanks* (Ellis *et al.*, 1985).

The costs of long-term seed storage can be reduced

Experience has shown that in climates with high ambient temperatures the operation of large long-term seed stores at the preferred temperature of − 18 °C requires much power. Frequently the resources of genebanks are insufficient to meet the energy costs and conditions of seed storage slip below the recommended standards. Two possibilities of low-input storage have emerged recently.

Ultra-low seed moisture content may substitute for low temperatures

There is recent evidence that, for some species, particularly those with high oil contents in the seed, careful pre-storage drying to ultra-low moisture contents (below the conventional minimum of 5 per cent), can substantially increase seed longevity. There is a minimum moisture content below which no further gains in longevity can accrue and current research is determining what this is for a wider range of species.

In the case of *Brassica napus* (swede), viability increased twelve times when seeds were stored at 3 per cent instead of 5 per cent moisture, and a reduction from 5 per cent to 2 per cent gave an increase in longevity similar to a reduction of storage temperature from $+20\,°C$ to $-10\,°C$. The prospects, therefore, for trading-off ultra-low moisture contents for higher storage temperatures to achieve the same seed longevity are very encouraging. The substitution of refrigerator temperatures for the more usual deep-freeze temperatures would provide a considerable saving in storage costs.

One natural resource can be used to conserve another

An even more radical approach to simplifying and cheapening long-term storage is the use of locations where the ambient temperature is at sub-zero levels. The five Nordic countries, Denmark, Finland, Iceland, Norway and Sweden, jointly operate a Nordic Genebank through the Nordic Council of Ministers. They have decided to utilise as a back-up to their long-term Base storage, a gallery in a disused mine on the island of Svalbard, at a latitude of 80°N, well inside the Arctic Circle. Temperatures inside the mine are $-4\,°C$, plus or minus 0.1 °C, throughout the year, higher than the ideal but acceptable because of its low cost and high security and the possibility of compensating with storage at ultra-low moisture contents. The administration of Svalbard has been ceded to Norway by a treaty, guaranteed by more than fifty nations. The island is demilitarised and to some degree may be regarded as international territory. The Government of Norway has offered to build a special facility at Svalbard to accommodate Base storage material from other countries and from the International Agricultural Research Centers. The operation of the storage facility would be similar to a bank safe-deposit and countries using it would not lose control of the germplasm which they deposited there.

The ambient temperature in the mine is higher than that currently recommended for long-term storage, but it may be that by the use of ultra-low moisture contents, as discussed above, acceptable storage times may be achieved. The very high security and no-cost storage will doubtless lead many countries to accept this attractive possibility, particularly if responsibility for its management is vested in FAO, a UN agency.

Some species cannot be stored as seed

Most of the major crop species are amenable to the convenient and relatively inexpensive method of conservation by *ex-situ* seed storage. A few, are not. Their seeds are said to be recalcitrant, a term which refers to their inability to retain viability after dehydration and low-temperature storage. They are usually large seeds with high moisture contents (cocoa, coconut, rubber, and many tropical tree fruits such as avocado, mango, mangosteen and jackfruit). Seeds of some recalcitrant species can only be stored, without dessication, for a few days, weeks or months (Roberts *et al.*, 1984).

Other crops are sexually sterile and are dependent on vegetative propagation for survival; examples are cultivated bananas, yams and sweet potatoes.

Species of these two types can only be conserved in the vegetative state. There are two general ways of doing this. The conventional method is to grow specimens in field or nursery plantings – Field genebanks, but, more recently, attempts have been made to devise methods for maintaining small pieces of tissue from the plants by *in-vitro* culture methods.

Field genebanks: a necessity for some species

Field genebanks suffer from a number of disadvantages. They usually require significant areas of land, if they are to contain adequate samples of the genetic variation of the species and particularly if the species is a tree crop. They require constant attention to control pests and diseases. It is common for vegetatively propagated species to be plagued with virus diseases which can seriously affect their survival. The history of the Commonwealth Potato Collection (CPC) provides a good example of this hazard. Much of the original collected material was brought from South America as tubers and for a number of years the collection was maintained as tuber lines planted and harvested each year. This system had clear advantages for potato breeders, who were able to select potential parents on the attributes of the clones which were available for inspection each year. However, as virus infections, often multiple infections, spread through the collection, the appearance of the clones changed and their viability decreased until, in order to save the material from extinction, clonal maintenance was replaced by maintenance as 'true' seed. 'True' seed is seed produced by the normal sexual process, in contrast to seed-tubers which are used for vegetative plantings. With the exception of a few RNA viroids, viruses are not transmitted through 'true' seed. The use of 'true' seed has proved to be a safe method of conservation of the genetic resources of the potato, but at a price. The price has been a loss of convenience to the breeder. 'True' seed progeny differ genetically from the well-described parental clone and from each other and breeders must grow and study them, usually for at least one year, before they can decide on which to use as parents.

In the case of sterile species and those with recalcitrant seed, this option is not available. Developments in recent years in the tissue culture of plants offer an alternative.

In-vitro *genebanks offer greater security*

The growth of plant cells and tissues in sterile culture is commonplace today. It is used widely as a means of vegetatively multiplying a range of agricultural, horticultural and ornamental species, a process known as micro-propagation. It has the merits of low cost and a high level of control of disease infection, although contamination is always a risk.

It is a characteristic of tissue culture that the nutrient medium stales with age and the cultured material needs to be transferred to fresh medium from time to time. When using tissue culture for germplasm conservation it is highly desirable to lengthen the interval between transfers. There are two reasons for this; first to reduce operating costs and second to reduce the possibility of genetic change in the conserved material. Genetic mutants which favour growth and survival under the culture conditions would tend to be selectively propagated at each sub-culturing cycle, thereby distorting the original allelic frequencies of the accession. Increasing subculture intervals can best be done by slowing down the rate of growth. There are two ways of doing this: by slow-growth culture, in which the cultures are grown at low temperatures and perhaps with chemical growth retardants, or, alternatively, by cryopreservation, where tissue is stored in liquid nitrogen, at a temperature of $-197\,°C$, when cell processes and growth come to a complete stop. Cryopreservation has to be followed by a regenerative culture phase in growth-stimulating conditions and may be regarded as a special kind of *in-vitro* storage.

In any form of *in-vitro* conservation the aim is to recover from culture viable material capable of regenerating into a normal plant which has the genetic characteristics of the material which was cultured at the outset. Each of the two methods has its own limitations.

The procedures used to obtain optimum slow-growth cultures depend on the species to be conserved (Withers, 1984, 1989). By choice of suitable procedures, sub-culturing intervals can be extended up to twelve or even twenty-four months, but whatever the interval, re-culturing eventually becomes necessary. The principal hazard in slow-growth systems lies in the likelihood of genetic changes being induced by the conditions under which the material is maintained. Slow-growth conditions are abnormal and therefore can be regarded as stress-inducing conditions. Mutations which favour survival under these severe and unnatural conditions are likely to be at a selective advantage, and the possibility for progressive genetic change over successive re-culturings cannot be ignored. Evidence is accumulating to indicate that under certain circumstances this is a common event (Scowcroft, 1984).

General statements about the potential of slow-growth tissue culture as an alternative to Field genebanks are not possible. The value of the technique is likely to vary from one species to another. A study of the operational feasibility of a tissue culture genebank – an *in-vitro* active genebank (IVAG), and of the genetic stability of the material in culture, is being made by IBPGR and the Centro Internacional de Agricultura Tropical (CIAT) using cassava as experimental material (IBPGR, 1990).

Cryopreservation offers the prospect of long-term storage without the need for periodic re-culturings and without the genetic changes associated with slow-growth tissue culture. At the temperature of liquid nitrogen, all biological processes cease and genetic stability is thought to be assured, although this remains to be experimentally verified. The maintenance of material requires little more than regular replenishment of the liquid nitrogen so that the long-term costs are likely to be low. However, the protocols for the preparation of the material for freezing and for recovery afterwards have yet to be worked out for most species.

Cryopreservation could have great potential for the storage of excised embryos of recalcitrant seeds but the species and tissue types for which it is an appropriate method of storage have yet to be determined. Much research remains to be done before this application of ultra-low temperature storage becomes a routine procedure. Additionally, it may provide an extra-secure option for the long-term storage of orthodox seeds as an alternative to the more usual temperature of $-18\,^{\circ}C$.

To sum up on the relative merits of the different possibilities of *ex-situ* conservation, it is perfectly clear that storage of seed at low moisture contents and low temperatures is the most effective and economical way of preserving genetic resources. For species with seed which is intolerant of dehydration, or for species which are sexually sterile, Field genebanks are the only current alternative, although slow-growth tissue culture and cryopreservation are possibilities for the near and mid-term future, respectively.

Ex-situ seed storage has its critics

From time to time, in discussions and in published articles, concern is expressed about the wisdom of conserving genetic resources *ex situ* as seed in genebanks. These criticisms of seed storage are made in spite of the many advantages of the method compared to alternatives; for example the natural suitability of orthodox seeds for long-term storage, the relative ease of controlling storage conditions and the high level of security and protection from pests and diseases. It seems appropriate to consider the nature and validity of these criticisms.

Are ancestral species all that is needed?

It has been suggested that the task of storing large numbers of accessions in cold stores and the associated activities of documentation, characterisation and seed multiplication, could all be considerably reduced by disregarding the landraces and primitive varieties and conserving only their ancestral species. It is argued that since the wild species gave rise to all the cultivated forms and since the wild species must be conserved anyway as the source of genetic diversity which is not available in cultivated types, then the ancestral species contain all the genetic diversity that we need to conserve.

In considering this argument, we may divide our crop plants into two groups. There are those which have been derived from a single ancestral species; for example the various root and leaf beets are all descended from *Beta maritima*. A return to *B. maritima* for genetic variation which might otherwise have been available in a primitive variety or landrace would add greatly to the time taken to achieve the required combination of new alleles from *B. maritima* with the otherwise desirable genetic background of the cultivated parent. Hybrids between modern varieties and wild species may contain the desired alleles from the wild parent, but one half of their total genetic constitution will also come from the wild species. The inevitable consequence is that the agricultural acceptability of the hybrid will be low. Usually, several to many generations of back-crossing to the agriculturally adapted parent are necessary, before a desirable combination can be found in the progeny. Breeders, when seeking new sources of variation, sensibly turn first to those which offer the prospect of most rapid progress towards their objective, that is to modern varieties, then to primitive varieties and, only when these fail, do they turn to the wild species.

The second group of crop plants consists of those which are derived from two or more ancestral species; bread wheat, for example, has three. Even if the ancestral species are known, and in some crop plants they are not, the retracing of the evolutionary pathway is likely to be a lengthy process of two stages. The first is to make the necessary sequence of hybridisations to produce a primitive form of the crop plant, often with problems of sexual sterility of hybrids to be overcome on the way, and the second, is the breeding and selection of agriculturally acceptable varieties from the synthesised primitive ancestor. These are formidable breeding tasks!

To abandon as an economy measure, the immense diversity of landraces and primitive varieties which have evolved in the development of the world's crops, would be perversely to complicate and lengthen the work of future crop improvement. A consequence would undoubtedly be to drastically reduce the use of genetic resources by plant breeders. There are more sensible and acceptable methods of reducing the demands on resources, for example in the rationalisation of collections, in the cooperative sharing of work and in the formation of Core collections (see pages 121–123).

Does ex-situ *seed conservation arrest evolution?*

An argument which is frequently raised against *ex-situ* conservation of seed is that it arrests evolution. Evolution, it is said, does not occur in genebanks. The implication is that had the seed been left in its natural habitat, either agricultural or wild, it would have continued to evolve into forms which would be different from those which were collected and stored. It is worth examining this idea, because of its popularity, first as it might apply to weedy and cultivated forms and secondly to wild species.

It is true that changes in farming practices, for example in sowing and harvesting times, in fertiliser use, or in the adoption of irrigation, could alter selection pressures and could therefore bring about changes in the genetic characteristics of landraces and primitive cultivars. It remains a moot point whether, if such changes had occurred, they would be advantageous or disadvantageous to the needs of posterity, but if they had, they would be contrary to the accepted aim of conservation; to preserve the natural genetic diversity which has evolved in the past.

Without doubt, the *in-situ* change most likely to occur is the total substitution of modern high-yielding varieties for the traditional ones. Experience shows that the old varieties could be completely eliminated in one or two years. It has been suggested that 'museum farms' or 'folk farms' are a way of overcoming this arrest of evolutionary change and are a preferable alternative to *ex-situ* conservation. On these farms, the traditional old varieties or landraces would be maintained under traditional systems of husbandry – the farm system would be held in a state of arrested development in order that the genetic characteristics of the crop populations could remain in a state of equilibrium with the 'historical' agro-ecological environment. Under these conditions, if they could be attained, the selection pressures from the environment would supposedly be constant, except for the normal variations in the climate, and therefore adaptive or evolutionary change would not be expected to occur. It seems, therefore, that the proposed 'folk farm' could not fulfil its stated purpose of ensuring the continuation of evolution; on the contrary, its function, if any, would be to prevent it.

This line of reasoning would seem to hold equally well for the particular case of the evolution of disease-resistance alleles. It is a feature of heterogeneous landraces and to a lesser extent of old varieties that the genetic heterogeneity extends to the alleles of disease-resistance genes – plants may be resistant to one or more races of a pathogen and susceptible to others. The consequence is that most diseases and most pathogen strains are to be found in most years, but at a low level, and therefore applying low selection pressures to the resistance alleles of the crop host. The outcome, an equilibrium of allele frequencies, is little different to the conservation of alleles that can be achieved by *ex-situ* conservation. It is difficult to see, therefore, how the preservation of landraces and old varieties in archaic but stable systems, can give rise to the evolution of novel resistance alleles.

For wild species, the issue is not the preservation of archaic farming systems, but rather the conservation of genetic resources *in situ* to permit adaptive or evolutionary change to occur. Again, we should ask, how would such change come about? The only known mechanism is by natural selection acting on the reproductive fitness of the individuals in the population. Selection pressures may vary somewhat from year to year in response to natural variations in climate, rainfall and temperature. But within the limits of these natural variations populations remain in a state of dynamic equilibrium with respect to the alleles of their genes.

Significant changes in allele frequencies can be expected if there is a progressive and directional shift in selection pressures, due to directional environmental changes. It is claimed that such changes are now occurring, but it is not clear whether any genetic changes they induce will be advantageous or not to plant breeders of the future, or whether alternatively they will bring about the destruction of the population or species.

Another cause of change in allelic frequencies is when catastrophic events destroy most of the population and leave only a small residual nucleus which, by chance, has allelic frequencies significantly different from those of the parent population, an event known as genetic drift. In these circumstances, loss of alleles, particularly rare alleles, can occur. In subsequent years, if selection pressures return to normal, the genetic structure of the population will approach its own maximum fitness and allele representation, and by virtue of 'founder effect', may differ initially from the original population. Lost alleles may eventually be recovered through mutation and migration from outside in cross-pollinating species.

If the aim is to conserve wild species in their natural environment, it is difficult to see by what means the expected and desired evolutionary changes are to come about. Long-term climatic change seems to be the only possibility, and we have no way of knowing whether the new allelic combinations which might evolve in response to this stimulus would be of any greater value to posterity than those available now.

Seed multiplication in genebanks can change allelic frequencies

A more realistic hazard of *ex-situ* conservation is that we may induce changes in gene frequencies of accessions when they are sown and grown-on for seed multiplication to meet the operating requirements of the genebank. Seed-renewal sowings are often made from small samples, under the environmental conditions pertaining to the area of the genebank. These conditions may be significantly different from, and therefore exert selection pressures different to, those which occurred in the natural habitat of the wild species or the farming system of the landrace. This very real problem will be discussed in Chapter 6.

Another practical difficulty which has to be faced in maintaining wild species *ex situ*, arises from the common occurrence of seed dormancy mechanisms, which have been largely eliminated in the domestication of most cultivated forms of major crops. Until methods have been devised for breaking dormancy at will, and this is not yet possible in all wild species (Ellis *et al.*, 1985), the dormancy of some seeds of a sample can distort the results of germination tests or cause the shifting of allelic frequencies during seed multiplication, because the seed-bearing plants will not constitute a random sample of the population.

These considerations lead to the conclusion that, if a species is not under threat of extinction or of serious erosion in its natural habitat, it is best to leave it where it is, and, having mapped its distribution range as part of a comprehensive study of the ancestral forms of a crop, to collect seed from the wild as required. If a species appears to be under threat, we have to assess the magnitude of the threat in terms of the size of the threatened area in relation to the distribution range of the particular ecotype. If the danger of loss appears to be real, then clearly rescue seed collecting and *ex-situ* storage are the only option and, whatever associated changes in gene frequencies may ensue, they are clearly preferable to total loss. The key factor in being able to make a considered sensible response, rather than a knee-jerk reaction, to apparent danger of genetic loss, is information on the distribution ranges of different ecotypes of the species.

Is the proper place to conserve genetic diversity in its natural habitat?

The idea of conserving wild species in the habitat in which they naturally exist – *in-situ* conservation – has the immediate appeal of rightness and simplicity. It is the obvious way to avoid the many problems which arise when seed of wild species is taken into *ex-situ* storage. For species in ecosystems which are not under threat of destruction, the most sensible and effective policy is to leave the material to conserve itself, in the wild, and any collecting which is done will be for utilisation, for research or breeding, and not for conservation. For those ecosystems which are in danger, positive actions are required if their survival is to be secured.

We can recognise two categories of endangered ecosystem. There are those where the destruction of the habitat is on such a vast scale that the risks and consequences are self-evident and the principle consideration is the urgency of remedial action. Some well-known examples are the remorseless destruction of the Amazon rain-forest, the deforestation of the Himalayan foothills and the desertification of the sub-Saharan rangelands. These are extreme examples which have caught public attention because of the biological, agricultural and social disasters which have ensued. Responses to these catastrophies are usually thought of in the wide context of nature conservation, but it should be pointed out that any conservation measure, if it is to be effective in the long term, should include consideration of genetic diversity, not only in relation to crop plant ancestors, but also in relation to the other living components of the ecosystem.

The second category of endangered ecosystems includes those cases which are smaller and less dramatic but not necessarily less significant to crop plants. In these cases, concern about the survival of particular species, the crop ancestors, may be the main driving force for action and genetic diversity the principal aim.

In such cases much data has to be collected and analysed before the location, size and number of reserves needed can be defined. Then legislation has to be enacted to ensure their establishment and security in perpetuity. It is useful to consider some of the issues which have to be faced and some of the preconditions which have to be met if the *in-situ* reserves are to fulfil their purpose of ensuring the long-term security of adequate genetic diversity of the wild ancestors of our major crop plants.

Where do the different wild species occur?

One of the basic requirements is to understand in some detail, the geographical distribution range and the ecological zones within this range in which different species occur. For many wild species, this information is either incomplete or totally lacking, yet it is essential for informed decisions on whether, for example, habitat destruction in a particular location requires conservation action. If the habitat supports a significant and unique amount of the existing genetic diversity of the species of interest and of its associated species in the community, then clearly urgent action is called for. If, on the other hand, similar communities in similar ecological niches are to be found elsewhere in the distribution range of the species, the pressure for action is clearly less. In either case, informed decisions depend on good data.

Data of this type are essential also, when determining the minimum areas of reserves, to ensure that they contain populations of adequate size to maintain intra- and inter-specific genetic diversity. The point may be illustrated by considering two contrasting examples. The ancestors of the temperate cereals are annual grasses, growing often as dense mixed populations such that areas of only a few hectares support millions of individuals. On the other hand, tropical tree crops such as mango or rubber may occur as widely dispersed trees in a mixed tropical forest. The specification of adequate areas, and the number of areas which are practicable, will differ greatly in each case, but again, in this as in all aspects of *in-situ* conservation, there is a need for good data on population numbers, allelic frequencies and breeding systems as a basis for effective action.

What kind of wild species should be conserved?

From the outset, a fundamental policy decision has to be made about which types of wild species should be the focus of *in-situ* conservation. We can recognise three general types: (a) the ancestors of those crops which are of major significance to world agriculture, (b) the ancestors of minor crops and (c) those species which are thought to have potential as new crops for the future. The latter category excites much interest at the present time, partly because some wild species are known to be sources of pharmaceutically important substances, for example vincristin, derived from the Madagascar periwinkle (*Vinca rosea*), and diosgenin, used in the manufacture of oral contraceptives, sex hormones and cortisone, and derived from some wild species of yam (*Dioscorea* spp.). Other wild yams are widely used in traditional medicines in Africa and Asia and yet others provide fish and arrow poisons (Purseglove, 1985c). In addition to these and other well-documented examples, there are thousands of species known to rural societies world-wide as medicinal plants, whose pharmacological function and biochemical activity is unknown or poorly understood. There is a huge task here to record indigenous knowledge and study the more valuable species in this important area.

There are many other cases of plant extracts which serve highly specialised functions to man and which are collected from the wild. These conceivably might be cultivated as crops in the future. The difficulty is that this category could include a large part of the plant kingdom, and that therefore predictive choices would have to be made about which species would be needed to provide new crops for the future.

It is instructive to note that despite the enormous unexploited potential in the plant kingdom, only four major crops have been domesticated and bred in the recent past: sugar beet, oil palm, edible oilseed rape and rubber. The conclusion to be drawn is that new crops succeed when they are developed to fill a clearly defined market opportunity. The failure of Triticale to become established as a major cereal may well be an example of this point.

The so-called minor crops, are minor on a world scale, but often of great significance locally. The need for their conservation has to be judged on a local or regional scale. Some examples are quinoa in the Andes, peach palm in Central and South America, durian in south-east Asia, datepalm in North Africa and south-west Asia and breadfruit in south-east Asia and the Pacific. Even if these are considered to be worthy of action, their conservation would be without any immediate prospect of use of the conserved material, for little if any breeding is done on these crops.

Even if we restrict our thinking to our first category, the ancestors of the major crops, we probably have to consider several hundred species. Some will have narrow, others wide, ecological and geographical distribution ranges and will be subject, in varying degrees, to genetic erosion. For practical purposes, this group alone, if fully investigated, represents more than can be attempted in the foreseeable future. Conservation of diversity *in situ*, therefore, must necessarily be selective and should be seen as complementary to *ex-situ* conservation of seed, to Field genebanks and to *in-vitro* methods, not as an alternative strategy.

Collaboration between conservation bodies is necessary

Plant communities often contain ancestral species of more than one crop plant. For example, wild grass ancestors of wheat, oats and barley, of fodder grasses and of *Prunus* species are found together in ecosystems in Asia Minor; Brazil nut, passion fruits, *Papaya* and rubber in New World tropical rain-forest, and mango and citrus in Old World rain-forest. The aim of *in-situ* conservation, should be to choose sites which maximise the conservation of diverse populations of as many target species as possible.

Many organisations are active in the field of conservation of natural habitats. Their interests usually differ and few have any direct concern in the conservation of crop genetic resources. Conservation areas, national parks and general nature reserves are multiplying rapidly, particularly in developed countries, and if the conservation movement is to achieve and retain credibility in the minds of the general public, it must be capable of moderating its diverse enthusiasms with common sense, of operating in a collaborative manner as far as is possible, and of showing at least some awareness of the need to be cost-effective. The establishment of *in-situ* reserves is a conservation activity where there is clearly a need to integrate the efforts of national and international bodies for the conservation of their various target species – crop ancestors, ornamentals, drug sources, browse, fuelwood and timber species and, of course, many animal species. Carefully sited reserves can meet the needs of more than one conservation project.

In-situ *reserves require positive management*

When decisions have been taken on the best available scientific evidence and the boundaries defined, the reserves must be securely established by appropriate legislation. This may be by national governments acting alone or, if species distribution ranges cross national frontiers, cooperatively, or, in cases where a network of reserves is necessary in an integrated programme, by multilateral action. The definition of reserves is only the first step. Provision of continuing financial support and trained personnel for the management and use of the reserves is equally necessary, though often extremely difficult to find in parts of the world where they are needed most. If, as is frequently said, crop genetic resources are the heritage of mankind, then it is difficult to reject the corollary that the responsibilities for their conservation are also those of mankind.

The question arises, what kind of management is needed for an *in-situ* reserve? The answer has two parts. The first has to do with security. Designated areas have to be protected from agricultural and industrial development and from commercial exploitation – logging is an obvious example. But these areas have also to be protected from undue pressures from the local population, perhaps of subsistence farmers, who in times of drought may lack grazing for their stock and fuel for cooking. Reserves will frequently be needed in areas where human population increase and pressure on land resources is greatest. High fences and armed guards are impracticable and morally indefensible and the only possibility of maintaining a reserve is with the willing cooperation of the local people. Legislative controls need to be supplemented by public understanding of, and sympathy with, the purpose of the reserve. If this is lacking, then it is difficult to see how the future of a reserve can be assured.

The best prospects for long-term security are likely to be when the reserve is integrated into the life and agricultural economy of the surrounding area, but, it has been suggested, separated from it by a designated buffer zone. This is the second and more complex part of reserve management and the more constructive one from the point of view of long-term security, for in so far as a reserve becomes an accepted part of the agricultural and social system, it ceases to be an isolated target for developers and exploiters. The achievement of this degree of public acceptance of the objectives and methods of *in-situ* conservation and its integration into the farming system of the surrounding area, will be difficult but essential. The alternative, of imposing conservation reserves by government decree, offers the prospect of an endless protective vigil against deprived and unsympathetic graziers or fuel gatherers.

We can see then that the attractive simplicity of conservation in the wild is deceptive. It is rather more than just erecting a fence around an area of interest. Many scientific, economic and social problems have to be solved before *in-situ* conservation can be used with confidence for safeguarding genetic diversity. It undoubtedly offers, in the long term, the most satisfactory way of conserving wild species but, because of the length of time required to set up a reserve, it provides a solution for the future but rarely a means of rescuing the genetic resources which are under threat now. *In-situ* conservation is more suited to the prevention of genetic erosion than to limiting damage which has already begun. Present emergencies can only be dealt with by urgent rescue collecting and conservation *ex situ* which brings us back again to the point that *in-situ* and the various methods of *ex-situ* conservation, should be seen as complementary ways of conserving crop genetic resources.

Monitoring genetic erosion

The need for rescue collecting, which has had such a dominant influence on collecting strategies in the past fifteen years, is likely to continue for some crops for the foreseeable future, although happily on a much reduced scale. Nevertheless, if national programmes and international coordinating bodies, such as IBPGR, are to respond adequately to emergencies, they need to know of looming dangers before the losses occur, rather than afterwards. The rate of loss can be more rapid in the case of the supplanting of landraces or old cultivars by new uniform high-yielding varieties, than in the case of loss of genetic diversity in the wild. The latter is usually the result of either natural events, such as climatic change, or to human development projects involving extensive change in land use. Wholesale changes in the Mediterranean littoral ecosystems due to industrial and tourist developments during the past forty years, are an example. Both natural and man-made changes to natural habitats usually occur over a longer time-scale than the simple replacement of old varieties by new ones, and therefore give more opportunity for response. In all cases there is a clear need for an early-warning system so that effective rescue collecting can be organised.

The IBPGR has devised a pilot scheme, called Genetic Resources Early Warning System (GREWS), which is being tested and refined by its field officers at the present time. The scheme seeks to quantify on a series of arbitrary scales the magnitudes of any particular threats, as a basis for the rational planning of responses and for the assigning of priorities should this be necessary.

Some of the major causes of genetic erosion which would have to be considered in such a scheme are: long-term climatic change leading to desertification; uncontrolled logging leading to deforestation; migration and resettlement of human populations; expansion of grazing lands and overgrazing; agricultural development activities including irrigation schemes; expansion of cultivation into new areas; introduction of modern high-yielding varieties and new cultural practices; and, finally, natural disasters – such as floods and volcanic activity – and civil disturbances. The significance of this last category of sudden and unpredictable events is chiefly in relation to the security of collections already made. Having identified a cause of probable erosion, it is necessary to assess the significance of the cause under, say, five criteria and using numerical scores of the magnitude of each. Such criteria are:

Time-scale (relative to cause): might range from 5 (urgent/immediate) to 1 (probable/long term).

Proximity to an area of diversity: from 5 (central) to 1 (marginal).

Uniqueness of threatened area of diversity: from 5 (only known location for these crop varieties or ancestral species) to 1 (commonly found in similar habitats in the surrounding area).

Amount of material already collected from similar habitats: from 5 (adequate) to 1 (none).

Condition of collected seed in store: from 5 (excellent) to 1 (poor).

Undoubtedly, criteria and scoring classes will change in the light of experience, but it seems clear that such a scheme, or similar, will be needed to enable the joint assessment of early warnings from different people in different parts of the world, and to enable priorities to be assigned in a rational way. The success of this innovative approach to the problem of objectively quantifying risk of genetic erosion and of assessing the urgency of the response required is dependent on the information reaching an operating agency such as IBPGR. Its field officers are constantly on the alert for information about impending threats to genetic diversity, but they each have enormous geographical areas of responsibility; for example, there are but two officers to cover the whole of North America, South America and the Caribbean. The system needs to be supplemented with additional sources of information and ways of ensuring action. The staffs of national plant genetic resources programmes could be drawn into a world-wide genetic erosion warning system.

Furthermore, IBPGR and national genebanks could strengthen existing links, and establish new ones with other international agencies which could supply pertinent information, for example with FAO, UNDP, UNEP, UNESCO, IUCN and WWF on the one hand, but also with the World

Bank, the EC, US-AID and similar bodies which organise large-scale development projects, and which can themselves be the cause of loss of genetic diversity. It is becoming more common for development proposals to include an environmental impact statement but there is a strong case to be made for the inclusion of genetic resource erosion assessment in every major development project.

6

Managing genetic diversity

Résumé

The proper care of genetic diversity, whether conserved *ex situ*, or *in situ*, requires continuous and detailed management, to ensure its security and to promote its use in breeding and research.

Management for security of *ex-situ* material begins with the correct handling of samples. For most crops and forages this starts with the cleaning and drying of seed in preparation for storage. The determination of the percentage viability of the sample at the time of its placement in cold store, and thereafter at appropriate time intervals, serves as a regular check on the well-being of the material. Viability testing inevitably consumes seed which may be in short supply and methods of testing are being researched to economise in its use.

When the viability of the accession drops below a critical level, (85 per cent has been proposed as an acceptable lower limit), it is necessary to regenerate it by producing fresh seed of high viability from plants derived from a sub-sample of the original seed. Careful management of the seed multiplication process is necessary in order to ensure that the genetic distinctness of the accessions is not compromised by uncontrolled cross-pollination, and hence gene exchange, and by the inadvertent mixing of seeds. Various procedures are used to control pollination. Choice depends primarily on the breeding system of the species; whether inbreeding or outbreeding, and in the latter case, on whether insect or wind pollinated.

For Field genebanks of perennial species, the maintenance of healthy plants is a major preoccupation and management practices to control or cure diseases, to eliminate pests and to propagate healthy stocks are of particular importance.

Standards of management of major Base collections are monitored by IBPGR on behalf of the international community.

Management as an aid to utilisation, centres around the establishment of databases on the accessions in store. Data are essential if users are to make informed choices of material for breeding or research, and they are of two general types: Passport data which describe the identity and origin of the accession and Characterisation data which define its essential attributes.

Routine activities of Active genebanks are resource intensive. Many collections are now dauntingly large, and their conservation to acceptable standards is often beyond the resources available. The formation of a representative sub-set of the collection, termed a Core collection, on which work may be more effectively concentrated, has been proposed and is under test. It is of potential benefit to both curators and breeders.

In addition, a separate tool to assist curators in the day-to-day management of their collections, is the Genetic Resources Accession Assessment Score (GRAAS). This also is under test and development. It is a practical attempt to introduce an objective numerical approach to the assessment of the quality of an accession and of its management.

There are two possible strategies open to each curator; to try to establish a fully comprehensive collection and to accept the considerable costs associated with its management, or, given the principle of free availability of genetic resources, to regard a collection as part of a dispersed world collection, and to collaborate with other curators in a network of crop collections, sharing work, responsibilities and benefits. The latter is generally seen as the only sensible way forward and world crop germplasm networks are currently under development.

Why should management be necessary?

The suggestion that high-quality, dry seeds in cold storage should need careful management in perpetuity, may evoke expressions of incredulity. What could possibly go wrong, with seed samples so carefully stored? The answer is that seeds are living plants, although in embryonic form, and can all too readily die. In doing so they can alter the genetic characteristics of the sample to which they belonged. Although pre-storage drying to less than 10 per cent moisture prevents them from germinating, and low-temperature storage slows down cellular processes, so that the seeds can remain alive for longer, the point comes for all accessions, though at different times for different species, when living processes begin to decay and viability begins to drop as shown by an increasing percentage of the seeds of a sample failing to germinate. Viability is measured as the percentage of seeds which germinate – develop a visible root system – in a standard period of time, after a random sample taken from the accession has been set up in conditions which favour germination. Some seeds die soon after storage has begun, others after a long time, and while this differential dying can be due in part to differences in the pre-storage environment or management of the seed, it undoubtedly will be partly due to genetic factors and perhaps to other causes not yet identified. Survival of the plant in its seed phase is certain to be under genetic control, because of its critical importance to the subsequent development of the plant and the survival of the species. However, there is little direct evidence on this point (Roberts, 1975).

If the death of seeds occurred independently of any other genetically determined attribute, then it would not have any consequences other than for the seed which died. However, genes which influence length of storage life, like other genes in the nucleus, are not inherited independently but occur in associated groups which are linked together on the same chromosome. Therefore, to the extent that early-death alleles are linked with combinations of alleles of other genes, the death of some seeds could disturb the frequencies of the alleles of these other genes in the seed sample. In this way seed death not only reduces seed numbers but more importantly can change the genetic structure of the population. The aim of conservation is to preserve the original genetic characteristics of the accession; hence storage methods must minimise change.

Long storage life also brings its own hazards and these have been extensively studied. It is known that interactions of time, temperature, seed moisture content and oxygen availability, lead to predictable losses of seed viability and to genetic change in the surviving seed. The genetic changes occur as chromosome damage and as gene mutations (Roberts, 1975). As seed mortality increases, so also genetic damage increases in the survivors. Doubtless, the same correlation occurs in populations of dormant seeds in the wild, but there, deleterious genetic changes will be subject to the action of natural selection. Less-fit genotypes tend to be eliminated and the adaptation of specific ecotypes will be maintained. This is a self-correcting process which would not operate, or not to the same degree, under the conditions of seed multiplication in a genebank.

Because of these two inescapable consequences of seed storage – differential death and genetic damage – it is clearly important to renew stored seed stocks by seed multiplication before viability levels drop significantly. The IBPGR 'rule-of-thumb' is that the critical level is 85 per cent viability. This holds for many crops, especially cereals. It is expected, however, that crop specialists will agree their own regeneration standards for each crop; in some cases the minimum per cent viability will be lower and in others it may be higher.

A second reason why active management is necessary is that stored seeds are not museum material but are intended for use by plant breeders and other scientists. If they are to be used, curators have to be able to tell breeders what they have in their genebanks, where it comes from and what are its characteristics. Without this guide to the origin and nature of the seed samples, breeders would be at a loss to know where to begin in selecting material for use. Curators have to study the individual accessions to determine their significant characteristics and must record the information in a readily accessible form.

If genetic resources are to be used, three criteria must be met:

1. Seeds must germinate to produce plants which can be used as parents for crossing in a breeding programme.
2. There must be information on the characteristics of each accession, so that breeders may make informed choices.
3. The information must be readily accessible; computerised data bases are essential.

If a genebank is to meet these requirements for security and utilisation, curators have to employ active rather than passive management, and in this chapter we describe some management activities which are essential for secure and effective collections.

Management for security

There is no doubt that large quantities of seed material which have been collected in the past and deposited in genebanks have been lost through neglect. Landraces and old varieties which should have been safe for posterity have died in store for want of adequate care. In most cases, this was because seed collections were not kept under the correct conditions. The genebank as we know it now, with careful seed drying and low-temperature storage, is a development of the past twenty years. Prior to this time, most collections were stored under ambient conditions in paper bags in filing cabinets or on laboratory shelves. Many workers would have regarded the loss of samples through seed death as an inconvenience which could be rectified by recollecting. Today this remedial action is not an available option in the case of much cultivated material, and we have to make the most of what has already been collected. What are the essentials of good management for ensuring the survival of accessions?

Seed must be in good condition when it enters cold storage

Most collecting in the field is conducted under heavy pressure to cover as much ground and collect as much material as possible. The essential operations of seed collecting, packaging, labelling, recording of data on the habitat, plant community or farming system, as well as the pressing of plant voucher specimens to provide a reference to the characteristics of the mature plant from which the seed was collected, all have to be done quickly in order that as many sites as possible may be visited within the seed-ripening period of the species. For cultivated forms, there is the alternative option of collecting from farmers' stores or from markets, but for wild species the only alternative to collecting seeds from the plant is picking them off the soil surface or out of fissures after they have been shed. In consequence, collectors tend to concentrate on those things which can only be done in the field and to leave other work until the material is back at the genebank.

At the genebank a typical sequence of operations begins with the extraction of seeds from fruits, or from inflorescences as appropriate. The seed is cleaned, usually by winnowing, to remove empty seeds, weed seeds and chaff, and seed numbers are determined. After careful drying, seeds are sealed in moisture-proof containers and moved into cold storage. All the data obtained in these operations on seed weight, number and moisture content is recorded for each accession in the management-information file and to this should be added viability data from the first germination test. Assuming that enough seed has been collected, some is deposited for long-term storage in a Base collection with a security duplicate in another Base collection elsewhere and some goes to the medium-term store of an Active collection.

Individual seeds in a sample frequently differ in their maturity at the time of collection. Plants from heterogeneous landraces can differ in maturity dates even though ripening may be uniform on an individual plant. In other cases there may be sequential ripening within and between infloresences on the same plant, for example in forage grasses and legumes and many wild species. The premature drying of seeds brought about by their collection when immature can have adverse effects on viability and a germination test is the best way of measuring this. The results permit a prediction to be made of the date on which the next viability monitoring test should be done.

Discussion of the principles and details of the various procedures involved in these operations are to be found in Ellis *et al.* (1985).

Routine viability testing is necessary from time to time

Even well-prepared seed, stored under good conditions, will eventually decline in viability. Routine testing is the most reliable way of determining when the time has come to replace the stock with fresh seed, produced from plants derived from the genebank sample. This should be done when the viability drops below the recommended level for the species. Some curators consider the level of 85 per cent recommended by IBPGR for most annual crops, to be unattainable under their operating circumstances and others consider that it is unnecessarily high. An important consideration here is the point already discussed above, that, if the viability drops too far, then the sample of seed from which the regeneration plants are derived will not be a genetically representative sample of the original population. On the other hand, the act of regeneration from a random sub-sample and probably under environmental conditions different from those under which the original population evolved and grew, is itself liable to induce genetic changes. The greater the number of regeneration cycles, the greater the probability of genetic change. A curator, therefore, must strike a balance between the hazards of low viability standards on the one hand and too frequent regeneration on the other.

Species differ greatly in the inherent storage life of their seeds. Seed of some species found in archaeological sites have been dated by association with the site and then germinated (Harrington, 1970; Odum, 1965). The data include few cultivated species, but timothy (*Phleum pratense*), smooth stalked meadow grass (*Poa pratensis*), alsike clover (*Trifolium hybridum*), red clover (*T. pratense*) and white clover (*T. repens*), germinated after 21, 39, 30, 41 and an amazing 600 years, respectively. *Lupinus arcticus* is credited with germination after 10 000 years! More credible, perhaps, is the case of seeds of lotus, *Nelumbo nucifera*, unearthed in 1952 in the Liaodong Peninsula, China, and dated by carbon-14 techniques as being 950 ± 80 years old. Plants derived from these seeds may be seen growing in the water-lily pond of the Beijing Botanic Garden.

These examples and the more comprehensive list quoted by Harrington, give impressive evidence of seed longevity in certain species. Experimental data on inherent differences in longevity among species is scarce, but it is clear that differences do occur (Roberts, 1975). As a general rule wild and weedy species possess greater seed longevity than improved cultivated varieties. Curators cannot therefore adopt a standard generation interval for all the accessions in their genebanks. Ideally, a separate decision will be made for each.

Seed dormancy is a natural impediment to the determination of seed viability because it prevents germination of viable seeds and gives a false indication of low viability. Although it is of little importance in highly selected annual crops such as the major cereals, it is common in forage grasses and legumes. In wild and weedy species it is clearly an adaptation favouring survival. The physiological mechanisms of dormancy differ from one species to another so that the means by which they may be overcome will also differ. The IBPGR *Handbook of Seed Technology for Genebanks* (Ellis *et al.*, 1985), provides information on the occurrence and nature of dormancy mechanisms in fifty-eight families of the plant kingdom, and on procedures for breaking the dormancy and conducting germination tests. For many wild and weedy species, curators will find published information no more than a guide and will have to refine procedures for use on their material if they are to avoid erroneous estimates of viability.

Good management consumes some of the stored seed

The inescapable fact is that good management, in the form of viability tests, uses up potentially valuable and perhaps scarce seed from the store. The usual procedure is to sow a random sample of the accession – often 400 seeds – in a specially constructed germination unit, where temperature and humidity can be controlled, and to record the number of seeds which germinates over a fixed period of time. The resulting seedlings with emerged roots are then discarded since they cannot be dried again. The temperatures used and the duration of the tests vary with the species (Ellis *et al.*, 1985).

Stanwood and Bass, quoted by Ellis and Roberts (1984), report that viability monitoring tests are the major cause of seed depletion within accessions at the National Seed Storage Laboratory, Fort Collins, USA (and this is almost certainly the case in many other genebanks). In one year, three million seeds were used for tests and only twenty thousand were sent to breeding programmes! A non-destructive viability test would be a major contribution to the conservation of stored seed and Fort Collins is currently carrying out research to this end.

There are three factors which affect the quantity of seed used in viability testing: sample size, testing frequency and the regeneration standard (the minimum acceptable germination percentage which, when reached, indicates the need for replacement of the accession sample by fresh seed of high viability). Tests are carried out on sub-samples and therefore give estimates of viability which are subject to error. The larger the sample the smaller the error and the lower the probability of regenerating an accession unnecessarily, or, alternatively, of failing to regenerate when needed.

From the outset, IBPGR has been concerned about the undue consumption of seed in routine viability testing, and, in consequence, the statistical problems associated with the prediction of true viabilities were examined by Ellis and Roberts (1984), to see if more 'economical' methods could be found. The standard procedure is to use a sample of fixed size, say 400 seeds. In order to economise on precious seeds, Ellis and Roberts suggest that tests of fixed sample size can be replaced by sequential tests where the number of seeds required is not predetermined. Their method involves testing groups of 40 seeds in sequence. If the number of seeds germinating in the first batch of 40 is equal to or less than 29 then that is taken as sufficient evidence for regenerating the accession with a 95 per cent probability of being correct when the regeneration standard is 85 per cent. If the number germinating is between 30 and 40, another sample of 40 should be tested and the number germinating should be added to that from the first test, giving a joint sample size of 80 seeds. If a total of 76 seeds or more have germinated, then the accession can be left in store with a probability of 95 per cent that its germination percentage is 85 or more. If, however, the number of seeds germinated lies between 65 and 75, the test should be continued with another 40 seeds. On average, a sequential test will require substantially fewer seeds and have equal reliability to a germination test based on a fixed sample size of 467 seeds.

Accessions with high and low germination percentages will be identified with small seed samples, and sample sizes will progressively increase for accessions of intermediate viabilities, particularly for those in the range 85–90 per cent, but only as far as is required to reach a decision. Curators interpret their results at each stage simply by reference to a table designed for the purpose (Table 6.1).

The most simple and direct way of economising in the use of seed for viability testing is to test less often. This is possible without deleterious results to the accession, providing that the seed was of high viability when put into store under the recommended conditions of low moisture and low temperature. An important factor in this aspect of management is the

Table 6.1. *Table for intepreting data from a sequential seed germination test, using successive samples of forty seeds. Regeneration is necessary when viability is estimated (with a probability of 95%), to be less than 85%. (Adapted from Ellis and Roberts, 1984)*

No. of seeds tested (Cumulative)	If number of seeds germinated is		
	Equal to or less than	Between	Equal to or more than
40	29	30–40	–
80	64	65–75	76
120	100	101–110	111
160	135	136–145	146
200	170	171–180	181
	Regenerate	Continue test as necessary	Continue storage

experience and judgement of the curator in knowing how long the testing may be delayed for a particular accession. In this connection, it is important to stress that the unit for action and for management decision is the individual accession. Attempts to use results from selected accessions as indicators of the viability status of a larger group to which the tested accessions may belong (the group may consist of accessions of the same species or of the same initial germination percentage), are unreliable and unacceptable short-cuts.

With regard to Active collections, the question of when to regenerate is often decided by the consumption of seed in routine genebank activities. Seeds are used up by distribution to breeders and researchers and by the sowing of samples for characterisation studies so that the replenishment of stocks by seed multiplication may be necessary before low viability requires it. In the course of routine activities which require the germination of seed, information on viability inevitably accrues, and this may be used as an indicator of the status of corresponding material in Base collections.

Seed multiplication: necessary but hazardous

There are three principal reasons why it is necessary to renew seed stocks of accessions. First, the collected sample of seed may be too small to meet the requirements of Active, Base and Duplicate collections, a not uncommon occurrence with wild species material which is often difficult to collect in quantity. Secondly, existing seed stocks may have reached a low level, in Active collections, because of normal genebank activities such as distribution of seed to breeders and researchers, or use in characterisation studies. The more active the genebank, the more frequently will renewal be required. Seed renewal is also required when viability drops below the acceptable level. The second and third circumstances apply mostly to Active collections and only rarely to Base collections, which are seldom used and which are stored at lower temperatures.

The process of seed renewal is a superficially simple matter of sowing some of the surviving seed from the accession in store, growing the plants to maturity in appropriate environmental conditions, harvesting the seed produced and storing it in place of the sample which had lost viability. Complications can arise from the way in which the seed is produced.

The reason for conserving many accessions of the same crop or species is that they are genetically different from each other. It is the genetic differences between accessions which is so important, and seed renewal through multiplication can represent a serious threat to the distinctness and genetic integrity of the accessions being multiplied. The threat derives from a number of different hazards, and in this section we will assess their significance and discuss what can be done to minimise their effects.

The breeding system of the species determines the procedures of seed multiplication

We have discussed in Chapter 1, the central role played by sexual reproduction – the exchange of genetic material between individuals – in the generation of genetic diversity and how this diversity, subjected to the powerful effects of selection, either natural or artificial, leads to the establishment of allelic combinations which confer adaptation to particular environments. Plants of most species regularly exchange genetic material by means of cross-pollination and many have evolved genetic mechanisms which ensure that this happens. Cross-pollinating species are said to be outbreeders.

Several major grain crops, for example wheat, barley, rice, oats and beans, under the influence of man during domestication and perhaps in response to selection for uniformity, have changed their breeding system !rom outbreeding to inbreeding. They self-fertilise to a variable but usually high degree, often 95 per cent or more. The wild ancestral species of these crops, however, retain their outbreeding habit and most wild species and

Table 6.2. *Breeding systems of some crop plants*

Self-fertilising inbreds	Cross-fertilising outbreds	
Annuals	Annuals/biennials	Perennials
Wheat	Maize	Forage grasses
Barley	Rye	Forage legumes
Oats	Brassicas	Grape
Beans	Beet	Hop
Peas	Onions	Hemp
Tobacco		Apple
Tomato		Pear
Cotton		Plum
Flax		Date
Grain sorghums		Mango
		Tea
		Coffee
		Cocoa
		Oil palm
		Coconut

most crop plants are outbreeders. A small number of crop plants have an intermediate breeding system; for example the broad bean (*Vicia faba*) usually has about a third of the flowers self-pollinated and two-thirds cross-pollinated by bees. The breeding systems of some major crops are given in Table 6.2.

Knowledge of the breeding system is important for seed multiplication by genebanks, both in the mechanics of pollination and in its consequences to the genetic characteristics of the samples.

In outbreeding species, both individuals and populations are genetically highly heterogeneous. This is a consequence of the genetic mixing which cross-fertilisation brings about and which continues from generation to generation. Consequently, the identity of particular genotypes cannot be maintained unchanged but the average expression of a character in a population can remain in equilibrium. Maintaining these mean expressions from generation to generation, and hence the genetic integrity of the accession, is a matter of maintaining unchanged, the frequency of the alleles in the accession. What this means in practice, is that the pollen and the eggs which give rise to the next generation, should collectively transmit the alleles in the same frequency as that in which they occur in the parental sample.

In inbreeding species individuals are characteristically highly homozygous – the two alleles of each gene are usually identical – but populations of inbreeding landraces can be highly heterogeneous – individuals differ from each other. Variation, which in outbreeders is partly concealed within the genetic structure of individuals, is revealed in inbreeders as visible differences among individuals; for example, landraces of inbreeding bread wheat are much more diverse in their external characteristics than landraces of outbreeding rye. Because self-fertilisers are homozygous, or largely so, they give progeny which are genetically and in appearance, similar to themselves. This is not wholly true, because some cross-fertilisation occurs in all inbreeders, although the frequency is often less than 1 per cent. In inbreeders, therefore, it is important when selecting the seed sample for multiplication that the frequency of different seeds in the sample for sowing should reflect the frequency of the different homozygous genotypes in the landrace population from which the accession was originally drawn.

Another aspect of the breeding system which is important to the genetic integrity of the accessions under multiplication, is the pollination mechanism and the liability of plants to be contaminated by foreign pollen. In strict inbreeders, such as wheat and rice, the probability of this happening is low but variable, according to the genotypes which are being multiplied and according to the climatic conditions at the time of flowering. Cross-pollination is more likely in warm, dry climates than in cool, humid ones. Effective control of low-risk cross-pollination is usually achieved by careful attention to the location of plots in the seed multiplication nursery. The planting of different species in adjoining rows provides effective isolation because sexually incompatible species act as pollen traps.

For outbreeding species, control of pollination is the major issue in the maintenance of genetic integrity. Many species are wind pollinated, most grasses including the wild ancestors of the major cereals, for example. Others, like the forage and grain legumes, are insect pollinated and the insect species involved are many and varied. Effective cross-pollination can occur over large distances. Consequently the methods of isolation differ according to the species but in general they fall under three different heads: spatial or temporal separation, natural or artificial barriers and hand-crossing accompanied by bagging of the inflorescences (Breese, 1989).

Spatial separation is of course influenced by the direction of the prevailing wind but may require distances of tens of metres between plots. Since, in any one year, a curator may have hundreds of accessions to multiply, it may well prove difficult to find sufficient isolated sites; isolated not only from each other but also from sexually compatible crops and wild species in the vicinity. Hand crossing and bagging provide the surest but the most labour-intensive option and the use of isolation chambers cooled by filtered air, in specially constructed glasshouses, the most capital intensive. The planting of multiplication plots as islands in tall-growing crops, which act as barriers to pollen flow, is often the most feasible and effective way of achieving reproductive isolation.

The sample used for seed multiplication is of critical importance to the maintenance of genetic integrity

It follows from the discussion above, that the genetic composition of the seed sample used to raise the plants for seed multiplication, is of critical importance in ensuring that progeny populations resemble parent popula-tions and that differences between accessions are faithfully maintained. There are a number of factors which can operate to change the genetic characteristics of accessions, most if not all of which can be prevented or minimised by good management practices.

It is customary to use a small sub-sample of seed from an accession to produce a fresh bulk for return to the genebank. This sub-sample used for multiplication, constitutes a bottleneck in the genetic continuity of the accession. It has to be remembered that it is taken at random with regard to the genetic content of the seeds, because there is no way of determining the latter from the seeds' appearance. If the sample is below a certain critical size, there will be insufficient plants to carry forward a representative sample of the genetic diversity of the original accession. Therefore the genetic structure of the second-generation accession can differ from the first, simply because of the size of the sub-sample from which it grew. This type of change, which can occur in nature, when cataclysmic events leave only a few survivors of a population, is known as genetic drift of allelic frequencies.

Various workers have estimated critical minimum plant numbers for maintaining the genetic integrity of accessions at a specified level and at a given probability. For example, Marshall and Brown (1975), define the optimum sample size per collection site, as the number of plants required to obtain, with 95 per cent probability, all the alleles with a frequency in the population equal to or greater than 5 per cent. If genetic integrity is to be maintained through subsequent seed multiplication, then the same criteria should apply. Gale and Lawrence (1984) have shown the importance of maintaining this minimum population size through successive sowings, and that the introduction of a tight bottleneck at an early multiplication cannot be corrected by use of overlarge sample size later. The theoretical calculations show that the minimum effective population size needs to be about fifty plants. This means that if each of the plants contributes equally both eggs and pollen to the succeeding generation, and if mating is completely at random, then fifty plants is enough to ensure the retention, with a probability of 95 per cent, of alleles with a frequency of 5 per cent or more in the accession.

The general assumption of random mating is probably wildly optimistic and rarely justified. There are significant differences between plants in the numbers of ovules and pollen grains which they produce, and furthermore, it is a common feature of outbreeding populations that plants reach reproductive maturity at different times, so that crossing tends to occur between those individuals whose sexual maturities coincide. Both of these factors reduce the effective breeding size of the population and hence increase the probability of drift. If minimum sample sizes are being grown, curators have to try to ensure equal genetic contributions from all plants. For this purpose, a number of pollinating systems are available, which vary in efficiency and which make different demands on genebank resources.

The various possibilities and their consequences are reviewed by Breese (1989) in a wide-ranging discussion of the problems of seed multiplication and the maintenance of the genetic integrity of accessions. It seems clear that, except for the labour-intensive system of crossing plants by hand, in pairs, and maintaining the progeny of the pair crosses separately, most of the feasible options would have effective sample sizes which are less than the number of plants used and therefore sample sizes ought to be at the upper end of the range of fifty to one hundred plants.

Seed multiplication provides many opportunities for unwanted changes due to selection

We have previously considered how heterogeneous populations of both inbreeders and outbreeders are maintained in a state of genetic equilibrium in their habitats by the action of selection, natural or artificial. Individuals with greater fitness tend to leave more progeny and genetic differentiation ensues between populations in different habitats. The genetic characteristics of each accession, therefore, are the result of the action of a unique combination of environmental factors peculiar to each habitat. Ideally, in order to maintain genetic integrity, a curator would need to reproduce, for each accession during multiplication, the combination of environmental conditions and hence selection pressures under which it evolved. It is, of course, manifestly impossible to do this.

Table 6.3. *Factors causing genetic shift during seed multiplication, and management practices to minimise their effects. (From Breese, 1989)*

Stage	Factors	Minimise by
Germination	Differential genotypic (i) longevities (ii) dormancy	(i) Regenerate before germination falls to <85% (ii) Artificially break dormancy
Seedling and vegetative stage	Differential genotypic survival due to: (i) interaction with climatic and soil factors (ii) susceptibility to diseases and pests (iii) competition	(i) Regenerate in habitat as similar as possible to that of origin of accession, or under controlled conditions (ii) Protect by fungicides, pesticides etc. (iii) Grow at low densities (i.e. spaced plants)
Reproductive phase	Differential production of flowers, pollen and seed	Maximise production from individual genotypes (see particularly (iii) above)
Harvesting, threshing, drying and packaging	Differential maturities and seed shedding	Harvest heads individually at appropriate stage of maturity. Bag heads to prevent loss of seed through shedding
Storage of high-quality seed	Differential maturities of seed may influence longevity	See above

There are two alternatives open to a curator. The first and most realistic is to grow the plants of the seed multiplication sample in such a way that it is subject to minimal selection pressures. If all plants produce seed and contribute seed in equal numbers to the following generation then the allelic frequencies characteristic of the accession can be expected to remain unchanged. The practical problems in achieving such a degree of control are formidable and have been summarised by Breese (1989) (Table 6.3). The factors listed in the table are both subtle and complex in their operation and it is unlikely that they can be negated completely by management, however skilled. The most that can be expected is that their effects are minimised by the procedures suggested.

The second alternative is to return the accession sample to the exact habitat from which it was derived, and to produce the seed there, where it will be subject to selection pressures which approximate to those that gave rise to it in the first place. This theoretically ideal solution is, however, quite impracticable and the nearest that one may approach it, is by means of cooperative multiplication agreements between genebanks in different geographical or ecological regions. Such arrangements are not only desirable, but essential in some instances. For example, CIMMYT cannot regenerate accessions of maize derived from the Andes nor can sorghum breeders in the USA satisfactorily regenerate tropical ecotypes. Many problems of these kinds have been solved by bilateral agreements between genebanks or national programmes. However, they only deal with a small part of the general need and the most promising general framework for such collaborative action is within crop diversity networks.

IBPGR monitors management standards in Base collections

Many national genebanks and a number in the International Agricultural Research Centers of the CGIAR have agreed with IBPGR to undertake the long-term storage of designated batches of accessions of certain crops. The terms and obligations of these arrangements are defined in letters-of-agreement in which the collaborating genebank agrees to maintain the accessions in accordance with the recommended international standards of the IBPGR and to make them available, without restriction, to curators of Active collections if their stocks of the accessions have been inadvertently lost.

From time to time scientific staff of IBPGR visit the genebanks, to check that the terms of the agreement are being observed. Sometimes, storage conditions are less than satisfactory, usually because of lack of resources rather than lack of will, and in these cases a report from IBPGR has been instrumental in providing the support which the curator had been lacking. By general consent this independent monitoring function on behalf of the international genetic resources community is useful and should continue.

Good management promotes the use of genetic resources

Good management depends on a good database

Seed is of little value without information describing its origin and characteristics. So a genebank has two equally important components; the seed in store and the database. The database contains all the data pertaining to each accession and is of critical importance to all other activities of the genebank.

Traditional methods of data recording were as manuscript notes in collectors' field notebooks, in genebank ledgers and in card index systems, all in formats personal to the collector or peculiar to the institution. This inevitably led to problems in data transfer from manuscript record to computer database, requiring skilled interpretation of the shorthand codes, symbols and conventions of earlier workers who had long since departed the scene.

There have been two important developments in the establishment of databases. The first was the development of computer programmes for the storage, interrogation and analysis of the data. The second is the development of common languages of description, in the form of descriptor lists for each crop or crop group. These lists were devised by groups of specialists. They provide a standard format for recording the data for each character (descriptor). For example, *Descriptors for Papaya* (IBPGR, 1988), provides for the description of the colour of the mature leaf stalk (descriptor 4.1.6), with descriptor states recorded on a 1–6 scale, where 1 = pale green and 5 = red–purple with intermediate values representing designated intermediate colours. Likewise, leaf shape (descriptor 4.1.7), may be described as one of six descriptor states by reference to standard leaf outlines (Fig. 6.1). Descriptor lists facilitate the recording of descriptive data in condensed form, and ensure its access and intelligibility to all who use the system.

One of the great advantages offered by computer databases is the possibility of interrogation. A curator can instruct the database to list all accessions which possess a combination of characteristics specified by a breeder, and by this means can respond rapidly with information on whether the genebank can meet the breeder's needs.

Passport data describe the identity and origin of an accession

Reference to any IBPGR descriptor list shows the same sequence of data items under the Passport heading. First there are the Accession data – including accession number, donor name (if the accession comes from another genebank), scientific name, acquisition date, date of last multiplication, number of times multiplied, number of seeds in accession etc. The second part of the Passport data consists of the Collection data, which are mostly concerned with where the accession comes from. They include, the collector's number and name, date of collection, country, province and map grid or other reference point to site of collection, latitude, longitude, altitude and ecological information (including associated plant species for wild species and agricultural data for cultivated varieties). An example of some typical Passport descriptors is reproduced in Fig. 6.2.

Many of the Passport data are provided to the curator by the collector or by the donor genebank. The curator's real work begins when he or she sets about the systematic description of the characteristics of each accession.

Characterisation data describe the essential features of each accession

Characterisation is concerned with the description of those characters which are sufficiently stable in their expression, irrespective of environmental influence, to typify the accession. This is straightforward in a species such as cultivated papaya, mentioned above, which has many stable pigment differences in male and female flowers, in fruit shape, in colour of skin and flesh of fruit, in seed colour and in leaf shape, although not necessarily so in a range of weedy related forms. Not all species with qualitative morphological differences which are suitable for use in characterisation present a straightforward task to the curator. Landraces of inbreeding wheat can be highly heterogeneous, with individuals differing from each other in a range of characters, and the characterisation of such a variable accession presents its own set of problems.

In mangosteen, another tropical fruit, and in many forage grasses there may be remarkable uniformity in the appearance of different ecotypes from across a wide distribution range. Where they do occur, differences are usually of degree and are influenced in their expression by the environment. The important temperate forage grass species *Lolium perenne* has few qualitatively expressed characters which can be used in the unambiguous identification of accessions. In this case characteristics may be inferred from Passport data describing its geographical, climatic or ecological origins or determined by labour-intensive replicated field trials and expressed as statistical means and variances.

It has been generally accepted that stable, environmentally independent characters should be included in Characterisation data and that characters of variable expression, such as plant height, seed number and ratio of leaf to stem should be classed as Evaluation data. The use of the term Evaluation in this context is unfortunate since it implies a judgement by the curator on the usefulness of the characters of the accession, and invokes a distinction in genetic expression which has nothing to do with value. Curators have the task of describing an accession – to characterise it – with as much objective data as possible, and may be assisted in the more specialised parts of this work by plant breeders and research scientists. Plant breeders need to make their own evaluation of an accession on the basis of the accession data and in relation to its suitability for inclusion in their breeding programmes. It is probable that the evaluation of an

4. PLANT DATA

4.1 VEGETATIVE

4.1.1 Tree habit

1   Single stem
2   Multiple stems

4.1.2 Number of nodes to first flower

4.1.3 Length of middle internode on tree [cm]

Mean of 5 measurements

4.1.4 Stem colour (adult trees)

1   Greenish or light grey
2   Greyish brown
3   Green and shades of red-purple (pink)
4   Red-purple (pink)
5   Other (specify)

4.1.5 Stem pigmentation

1   Only or mostly basal
2   Only or mostly lower
3   Only or mostly median
4   Only or mostly upper
5   Indiscriminate

4.1.6 Colour of mature leaf petiole

1   Pale green
2   Normal green
3   Dark green
4   Green and shades of red-purple
5   Red-purple
6   Other (specify in the NOTES descriptor, 11)

4.1.7 Leaf shape

See opposite

Fig. 6.1 Example of some characterisation descriptors and states. (Reproduced from *Descriptors for Papaya*, IBPGR, 1988.)

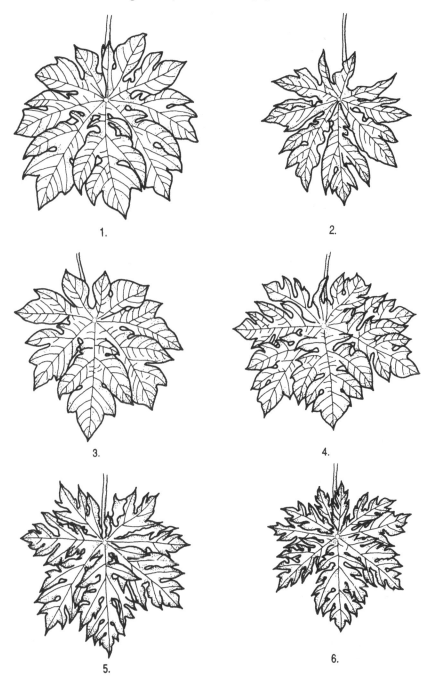

PASSPORT

1. ACCESSION DATA

1.1 ACCESSION NUMBER

This number serves as a unique identifier for accessions and is assigned by the curator when an accession is entered into his collection. Once assigned this number should never be reassigned to another accession in the collection. Even if an accession is lost, its assigned number is still not available for re-use. Letters should occur before the number to identify the genebank or national system (e.g. MG indicates that an accession comes from the genebank at Bari, Italy; PI indicates an accession within the USA system; ILL Indicates an accession in the ICARDA lentil collection)

1.2 DONOR NAME

Name of institution or individual responsible for donating the germplasm

1.3 DONOR IDENTIFICATION NUMBER

Number assigned to accession by the donor

1.4 OTHER NUMBERS ASSOCIATED WITH THE ACCESSION

(other numbers can be added as 1.4.3 etc.)

Any other identification number known to exist in other collections for this accession, e.g. USDA Plant Inventory number (not collection number, see 2.1)

1.4.1 Other number 1

1.4.2 Other number 2

1.5 SCIENTIFIC NAME

1.5.1 Genus

1.5.2 Species

1.5.3 Subspecies

1.6 PEDIGREE/CULTIVAR NAME

Nomenclature and designations assigned to breeder's material

Fig. 6.2 Some typical Passport descriptors and descriptor states (Reproduced from *Descriptors for Papaya*, IBPGR, 1988.)

accession by one breeder will differ from that of another, if their breeding aims differ.

Characterisation data has two uses. It is of importance to curators as a means of identifying the material in their charge (it tells them what the plants look like when they are grown from a seed sample and so enables them to check the authenticity of the accession), and it helps potential users to decide which of the accessions should be selected for their purposes.

Because the quantities of seed received at the genebank are often inadequate, curators, when growing plants for characterisation studies, often take the opportunity to arrange the sowings so as to simultaneously provide for seed increase.

This discussion of what ought to be done as part of good genebank management, should not obscure the sad fact that for many accessions in many genebanks it is not possible fully to implement all aspects of good management. Often Passport data are largely or completely lacking, either because they were not recorded when the accession was collected, or because seed samples have been exchanged between collections without the relevant data. This particularly applies to landraces collected long ago and which have ceased to exist outside of collections. In these cases, potential users are totally dependent on what information they can get from the Characterisation data produced by the curator and his or her collaborators.

It should be apparent from the foregoing discussion, that the work-load on curators and their staff in coping with the storage, databasing, characterisation, seed multiplication and viability testing of large numbers of accessions is considerable. Management of Active, as opposed to Base, collections is a complex business which requires considerable resources, often in excess of those available. Underfunding of activities in Active collections is in part due to the common misconception that when seed has been deposited in a genebank coldstore, the work is virtually over. In fact, it has hardly begun.

Core collections: a benefit to curators and breeders

A Core is a representative sub-set of accessions

Collections are often very large. Accession numbers of the major crops commonly run into thousands and not infrequently into tens of thousands. Size *per se* is not the important point, however, but rather, size in relation to available resources. Many genebanks have been set up simply as repositories for seeds and provided with staff and resources accordingly, without due regard for the intensive activities appropriate to an Active collection. Consequently, many collections have been swamped by an influx of accessions during the past ten to fifteen years and their curators have been overwhelmed, by the volume of work which the accessions generate and by their own inability to respond with the standard of management which they know to be necessary.

Thus, size of collections and inadequate resources have together become impediments to characterisation and documentation and hence to the utilisation of collected material. Without information to guide their choice, breeders and researchers are unable to discriminate among the multitude of accessions available. This regrettable situation in many genebanks is highlighted by some notable exceptions. For example, the rice collection at IRRI has some 83 000 accessions (Chang, 1989), but the curator and his staff are supported by a large team of scientific specialists who collaborate in extensive characterisation studies, particularly in the systematic screening for disease and insect pest resistances, critical characters for rice-breeding programmes (see Chapter 3).

For most collections, however, numbers are a problem and it is necessary to focus resources and activity in order to get started. The best way of doing this was first suggested by Frankel (1984), who proposed the formation of 'Core collections'. He defined a Core collection as a sample of accessions which would represent, 'with a minimum of repetitiveness, the genetic diversity of a crop species and its relatives'. Accessions not selected for inclusion in the Core would not be discarded but would continue to be stored as a Reserve collection.

The Core-collection concept enables curators to side-step some of the shortcomings in the composition of their collections arising from the often unplanned and uncoordinated acquisition of material during the years of urgent, sometimes frantic, rescue collecting. These shortcomings include redundant replication of the same material under the same or different names and uneven representation of accessions from different parts of the eco-geographic range of the crop or species. It follows that the Core would not be constructed from a random sample of the accessions but rather from carefully selected representatives from different ecological habitats or farming regions. Brown (1989) suggests that, on the basis of statistical considerations, a Core should consist of about 10 per cent of the accessions of the total collection and have a maximum of about 3000 accessions. He points out that because genetic diversity does not occur at random in a species, but is organised by the action of selection into functional groupings of alleles which confer adaptation to different habitats, carefully chosen accessions from distinct habitats will be efficient in retaining a disproportionately large fraction of the genetic variation of a species. The National Plant Germplasm System of the USA has been interested for a number of years in how best to form Cores in a number of its major collections.

A Core collection is a dynamic entity

It follows from the discussion above that the designation of certain accessions as a Core, cannot be the first step in genebank activities. If the necessary selectivity is to be exercised in its formation, the designation of the Core must be preceded by the registration and analysis of Passport data. When this has been done, the selection process may begin and when the first Core has been formed the activities of the genebank can be concentrated on it.

It is to be expected that the composition of the Core will change from time to time as opportunities occur for the curator to improve its representation of the genetic diversity of the species and in response to increasing knowledge accruing from the characterisation programme. For example, characterisation for disease resistances may reveal that certain eco-geographical areas are rich in resistance alleles and accordingly the representation of these areas may be increased by adding more accessions to the Core.

The weakness of the Core concept lies in the difficulty of implementing it in an optimal way. Lack of adequate Passport data and in many cases its total absence, can severely reduce the efficiency of selection of accessions for the Core. This difficulty is most common with the oldest material, often landrace samples, which are potentially the most valuable because of their antiquity and the great diversity which they contain.

The practical application of the Core concept to genebank management is currently under test and it seems clear that in under-resourced genebanks, which means most of them, it has much to offer as a means of using resources in the most effective way and of encouraging the use of stored genetic diversity by plant breeders. Its value can be further increased within the operational framework of emerging crop germplasm networks, an idea to be discussed below.

Large collections have their supporters

The case for large collections is argued with conviction by Chang (1989), the curator of one of the largest – the 83 000-accession rice collection at

Table 6.4. *Summary of insect-resistance screening tests at the International Rice Research Institute, up to 1984. (Modified from Chang, 1989)*

	Number of accessions		
Insect	Tested	Resistant	% resistant
(a) Cultivated rices			
Striped stem borer (*Chilo suppressalis*)	15 000	23	0.15
Yellow stem borer (*Scirpophaga incertulas*)	22 920	26	0.11
Whorl maggot (*Hydrellia philippina*)	16 918	1	0.01
Green leafhopper (*Nephotettix nigropictus*)	527	66	12.53
Green leafhopper (*Nephotettix malayanus*)	158	129	81.65
Zigzag leafhopper (*Recilla dorsalis*)	2 383	36	1.51
Leaffolder (*Cnaphalcrocis medinalis*)	20 816	117	0.56
Caseworm (*Nymphula depunctalis*)	5 183	0	0.00
Thrips (*Stenchaetothrips biformis*)	237	78	32.91
Rice bug (*Leptocorisa oratorius*)	406	0	0.00
Black bug (*Scotinophara coarctata*)	300	2	0.01
(b) Wild rices			
Brown plant hopper (*Nilaparvata lugens*)			
Biotype 1	446	204	45.7
Biotype 2	445	168	37.8
Biotype 3	448	178	39.7
Whitebacked plant hopper (*Sogatella furcifera*)	449	208	46.3
Green leafhopper (*Nephotettix virescens*)	447	239	53.4
Green leafhopper (*N. nigropictus*)	91	54	59.3
Green leafhopper (*N. malayanus*)	30	26	86.7
Zigzag leafhopper (*Recilla dorsalis*)	422	218	51.7
Striped stem borer (*Chilo suppressalis*)	243	13	5.3
Yellow stem borer (*Scirpophaga incertulas*)	322	70	21.7
Leaffolder (*Cnaphalocrocis medinalis*)	338	8	2.4
Whorl maggot (*Hydrellia philippina*)	339	7	2.1
Caseworm (*Nymphula depunctalis*)	304	0	0.0
Thrips (*Stenchaetothrips biformis*)	85	12	14.0

IRRI. Chang states, as a conclusion from extensive screening for disease and pest resistances, that the frequency of accessions of *Oryza sativa* varieties with resistance alleles is low and that only a large and diverse collection can be expected to contain them. It is difficult to escape the conclusion that some of the rare resistance alleles revealed by the systematic screening (Table 6.4), might have been excluded from even the most carefully selected Core collection.

Screening of wild species on the other hand, showed them to be a much richer source of resistance alleles. The extraordinarily comprehensive data presented by Chang (1989) (Table 6.4) indicate what can be achieved when there is adequate support for genebank activities and demonstrate the value of wild relatives as well as landraces in widening the genetic diversity of a collection and in making it more representative of the total genetic diversity of the crop gene-pool. Chang is the first to admit that he and his colleagues are highly favoured with a level of funding and scientific backing unlikely in a smaller collection and mostly lacking in other large collections. Nevertheless, smaller collections may expect to rise to these levels of achievement, by maximising their effectiveness through the use of the Core concept and through collaborative action (to be discussed below).

Genetic Resources Accession Assessment Score (GRAAS): a management tool for curators

The *ex-situ* conservation of genetic resources involves many diverse activities which have one common feature – they all generate data. If we imagine a genebank with say 5000 accessions and with only a modest fifty data items associated with each, the curator will have a quarter of a million data items to register, order, interrogate and analyse as part of his or her management activity. Success clearly depends on suitable computer programs as a principal management tool, but before the programs can be written, ideas must be crystallised and tested on what the management is aiming to achieve. One such aim is to be able to assess the quality of accessions, the completeness of the database and the standard of management of the collection material.

GRAAS is intended as a management tool to help curators to achieve these aims. It is, in part, a scoring system by which curators can assess the effectiveness of their management in relation to accepted international standards and which can provide a simple numerical assessment of the quality of their collection or of its components. GRAAS produces objective numerical indices of quality which can be used either as guides for management decisions for the allocation of resources into particular functions such as characterisation, or into particular parts of the collection such as wild relatives. The GRAAS data may also be used as objective evidence in support of submissions for additional resources necessary to raise standards to acceptable levels.

The basic idea is to assign scores according to the presence/absence of data for standard descriptors for the crop, or alternatively, according to the value of a descriptor state when a range of values is possible; for example, when scoring seed viabilities, percentages between 85 and 100 could score 9, 80–84 could score 5, 70–79 could score 3, and less than 70 score 1. Scales are arbitrary and will vary with the crop and the character being tested, and the opinions of the curator and will no doubt be modified by experience. Weightings can be introduced to accommodate differences in relative importance of the different characters.

The basic unit of data in the computing of GRAAS scores is always the numerical value assigned to a descriptor or descriptor state of an accession. From these units, overall scores can be computed according to need. For example, overall scores for individual accessions; for types of data – Passport, Characterisation, or Management data; for activities – viability testing or seed multiplication; or for parts of the collection – landraces or wild species. The idea is currently under test, modification and development in three major collections, of maize, groundnut and lettuce. The reactions of the curators concerned have been favourable.

GRAAS is one approach to raising standards of conservation of genetic resources in store. It is a management tool for curators which provides objective assessments of quality of conserved material, both for individual accessions and for the crop collection as a whole. It emphasises and measures the single most important feature of a genetic resource accession – its quality, as measured in both seed viability and essential information.

Crop germplasm networks

In the foregoing discussions on the collecting and management of genetic resources, we have referred to the benefits to be derived from collaboration between curators and national programmes within the framework of international crop germplasm networks. Crop networks are becoming more widely accepted as the key to effective genetic resources conservation and the solution to many of the problems facing curators working in isolation. We should consider: what is a crop network, what will it do, and what benefits will it bring to the conservation of crop genetic resources?

The starting point of any discussion on networks, is the assumption that the principle of free availability of genetic resources will continue to be observed. If it is, it follows that no individual collection needs to be comprehensive. Material not available from one collection will be obtainable from another. We can think of a dispersed world collection, consisting of the individual collections at the nodes of the network. Countries may choose to assemble and maintain a comprehensive collection, but the essential point is that within a collaborative network, none need do so and when the costs of managing to acceptable standards are fully appreciated, few are likely to reject the benefits of the collaborative option. One has to be realistic and recognise that there will always be political impediments to collaboration between some national programmes, but even here, a politically neutral international organisation such as a crop genetic resources network can provide a pathway for indirect collaboration.

The first task of a network would be to define what the world collection consists of, by determining, first within collection databases and then among them, which accessions are unique and which are unwittingly replicated, both within and among collections. Experience with the European barley collections has shown that this first step, even with computer facilities, is likely to be a formidable task for all major crops of which there are numerous large collections. A stepwise approach may be favoured, with the formation of regional databases which may later interact to form an international database for the crop. This may not be located in a single place, but may be dispersed like the genetic resources. Maize provides a good example; there is primary genetic diversity in the Americas, and secondary diversity in the Mediterranean, eastern Europe and also in parts of Asia, obvious locations for three or four regional databases.

When the existing world collection of unique material has been defined, other activities can follow. Inspired guesswork puts the frequency of replication of the same material in different collections as high as 50 per cent for many crops and exceptionally at 80 per cent in the case of barley, (J.T. Williams, 1989). It is likely, therefore, that curators will give high priority to the rationalising of their collections in relation to the composition of the others in the network. Redundant material may be consigned to a Reserve collection and work concentrated, by mutual agreement, on the unique material. The possibilities for reducing the unneccessary replication of work and for the more effective use of resources are enormous.

Analysis of Passport data can reveal gaps in the world collection in relation to the known eco-geographical range of the species or crop, and planned collecting organised to fill the known gaps.

Within the context of the unique accessions it is possible to apply the Core collection idea and to visualise the concentration of activities on a dispersed world Core collection. Prominent among these activities will undoubtedly be joint projects on the characterisation and seed multiplication of accessions. Both are demanding of resources and are therefore natural candidates for collaborative activity.

These are examples of the things which may be done in a network but it should be stressed that a network's programme will be determined by the collaborating curators and will undoubtedly vary from one crop to another. For a network to function most effectively, the curators of all of the major collections of the crop should take part, together with researchers actively working on problems related to genetic resources conservation.

By definition, the work of crop networks will be concentrated on the material in Active collections. However, an essential part of each network will be the associated Base collections of material for long-term storage and the collaborative framework provides the opportunity for ensuring that the resources stored in the dispersed Base collections represent the full range of the unique material in the Active collections.

The funding of this kind of collaborative activity presents unusual problems. First, there is the problem of inequality; inequality of inputs of unique material, of expertise and of resources, in cash and in kind. Wealthy nations will have to be prepared, for the common good, to assist the less wealthy, in addition to funding their own activities. Secondly, there is the question of continuity. Conservation is about making provision for posterity and there is no natural time limit which can be applied to crop network activities; at least, not at the moment. It is essential therefore that the funding of networks should be secure in perpetuity. This calls for commitments at the highest level between governments. Clearly, undertakings given by the directors of genebanks or of research institutes are insufficient since, as recent experiences in many countries show, funding at the institutional level is fraught with uncertainty.

It seems that nothing will do, short of an intergovernmental body, to which governments would make formal commitments to establishing and providing long-term operational support to plant genetic resources networks. The UN/FAO Commission on Plant Genetic Resources could be a suitable body through which governments could act to create and administer a fund to secure the long-term future of the networks and through which political issues relating to the operation of the networks could be resolved. IBPGR could perform the complementary function of providing technical and scientific support to the operation of the networks. Undoubtedly, crop networks are the most likely way of achieving the effective conservation and utilisation of crop genetic resources, and the first of them are now being established. The alternative – independent action – is now quite unacceptable and contrary to the rapidly developing recognition, that the solution of environmental crises and the conservation of the earth's resources are world problems which can only be solved by international cooperation.

7

Filling the gaps in the science base

Résumé

In coping with the flood of accessions into genebanks in the past fifteen years, curators have called upon established scientific knowledge as a basis for their conservation techniques and management practices. In doing so they have discovered that there are many gaps in the science base which need to be filled if crop germplasm is to be adequately collected, satisfactorily managed and effectively exploited in crop improvement.

Some examples of operational areas which require research are discussed:

The naming and identification of wild species relatives to many crops.
The collecting and transport of immature seed.
Genetic stability in seed stored for long periods and in material maintained *in vitro*.
The genetic cost of lowering storage and regeneration standards.
Breeding systems and the maintenance of genetic integrity during seed multiplication of many wild species.
Quarantine regulations and the distribution of germplasm.
The development of disease-resistance screening methods.

The need for an international body which has both an overall view and a detailed knowledge of what is needed is discussed. That role is currently filled by IBPGR which both stimulates and funds research, and it is argued that it should continue in this function and follow the principle of contracting-out projects to centres of relevant expertise. The new crop germplasm networks are expected to play a major advisory role on research needs.

Experience shows that the passive transfer of research results to the genetic resources community can be ineffective and active promotion of new ideas and methods may be necessary if problem-solving research is to fulfil its purpose.

Where are the gaps in the science base?

We have discussed earlier, how gaps occur in the representation in germplasm collections, of the diversity which exists in agriculture or in the wild. Of equal importance are the gaps in the background knowledge necessary to determine precisely what is lacking, how best to collect it and how to deal most effectively with that which has been collected. The phase of intense collecting of landraces and old varieties of the major crops (the minor ones remain to be done), which lasted about fifteen years, and the consequent flood of germplasm into genebanks, have revealed many practical problems whose solution depends on the filling of gaps in the science base. Furthermore, the recent switch of attention to the collection of germplasm of wild species has brought to light a new and different set of problems which cannot be resolved with present knowledge. These gaps occur in all aspects of genetic resources work; in collecting, conserving, characterising, regenerating and in germplasm distribution and exchange, and they must be filled if the operational standards are to reach acceptable levels.

By way of illustration, a few topics which are currently under investigation, or are in need of it, are outlined below.

Naming and recognising plants

The most pressing needs are often to do with the wild relatives of crops and their genetic diversity. What we need to know is often surprisingly basic. First, there is the fundamental requirement to know what the different species look like and how they are related to each other. Unless we can agree on what to call the different entities and on their descriptive characteristics – and disagreements on such matters are common, with different names being used for the same species and the same name for different species – the consequent confusion seriously hinders progress in conservation. So we need research on the taxonomy of many species groups.

Taxonomy is the branch of science which is concerned with the identification of organisms and with their classification into a hierarchy of groupings which as far as possible reflects natural relationships. There are different views among taxonomists about what constitutes a natural group. Classifications are usually based on the form of the plant in all its parts, but particularly on its floral parts, and on its habitat preferences. Not infrequently, different species are separated by no more than differences between the alleles of one or two genes. Such insignificant distinctions can lead to an unnecessary proliferation of taxa. From the point of view of genetic resources, which are intended for use in crop improvement through plant breeding, the ability of different forms to hybridise and produce fertile progeny, and hence to mingle their genetic material, is clearly an important criterion of relationship. Differences between such entities are unlikely to be permanent unless they are prevented from hybridising by isolating mechanisms, such as unbridgeable geographical separation or by different times of flowering. This approach to species definition, recognises as 'good' species those taxa which are reproductively isolated from each other. All to often, unfortunately, the genetic data necessary to apply this attractively simple criterion are not available.

Taxonomy is an old branch of botany and perhaps its best known. Its development and application enabled biologists to bring conceptual order to the apparently endless diversity of the living world. One of its earliest students was Carl Linnaeus, Professor of Botany in Uppsala in the 18th century. Despite the work of many taxonomists since Linnaeus and his contemporaries, the fact is that often it is still exceedingly difficult to identify some plants in the field, and this is so for many ancestral species of crop plants. Classifications are either lacking or, more frequently, are numerous, complex and conflicting. Research is needed to synthesise realistic classifications and to produce simple workable keys, which may be used for the unambiguous identification of material at the time of collection. Taxonomy, however, is now an unfashionable branch of botany and competent taxonomists willing to undertake this work are hard to find.

Simultaneously with the study of their taxonomy, it is necessary to determine the geographical distribution range of many wild crop ancestors, and their preferred ecological habitats. Much information is thought to lie in herbaria, attached to dried specimens collected in the past. As a first step, this information needs to be found, collated and analysed and then supplemented and extended by original research.

When plants can be identified and named and habitats and distribution ranges defined, collectors will be in a position to go out and assemble representative collections of the diversity existing in the wild. This multidisciplinary approach is frequently necessary in order to understand the orders of diversity and evolutionary pathways in a crop species group, as for example in the four crops discussed in Chapter 2, banana, bread wheat, cotton and rice.

Taxonomic confusion has been a major constraint on germplasm conservation in the genus *Allium*, and recently a Dutch, German and British collaborative project has been set up to clarify the taxonomy of those sections of this large genus which contain the edible forms; onions, leeks, garlic etc., and their wild relatives. The project, in addition to formal taxonomic studies in Gatersleben, Germany, and Kew, England, will include a survey by the Dutch collaborators of the species cultivated in south-east Asia and their relationships to the wild and cultivated species of southern China (IBPGR, 1990). The scale and scope of this project are indicative of the kind of work which needs to be done in many crop groups and of the multidisciplinary nature of much research in this field, involving, as it does, elements of taxonomy, genetics, evolution, plant geography and agriculture.

Another notable example of an attempt to bring order and clarity out of biological diversity is that of the International Legume Database and Information Service (ILDIS). Research groups in many countries are collaborating in the pooling of information on legume species diversity so that a world-wide information service can begin. The legumes are, of course, a large and diverse group of species, with world-wide distribution, which include many important forage and grain crops, shrubs valuable as browse and fuel and trees. All possess the enormously important attribute of forming symbiotic associations on their root systems with bacteria capable of fixing atmospheric nitrogen into compounds available to the host plant.

In the first phase of the database, it is proposed to compile a species checklist which will include accepted names of species, synonyms of the accepted names, and the place of the species in the classification of the family, together with data on the geographical distribution and ecological preferences of the species, vernacular names, economic importance and conservation status (whether common, rare or endangered). The checklist is being compiled on a regional basis, and work on the species of Africa, the Americas and Europe is in an advanced state. A beginning has been made on the legumes of the other continents. When complete, data on some 18 000 species are expected to be included. At a later date, it is proposed to add other modules of scientific information to the databases.

The intention is to disseminate the database widely to individual scientists and organisations in the form of compact discs (CD-ROM) and associated software and to link workers together with a world-wide information service.

ILDIS is very much an applied practical programme which will expand into many fields of science in relation to the legumes. It had its origins in problems of taxonomic confusion, and sorting out the names and characteristics of the different entities has been the first important step in facilitating communication between legume workers. Its strength lies in the large number of collaborating individuals and institutes from around the world who have specialist knowledge of the legumes. It, and the *Allium* example, are models of the ways in which groups of interested workers can collaborate to solve problems of common scientific interest. Regrettably, such initiatives have been rather rare, but it is expected that the crop germplasm networks now being established by IBPGR will provide a stimulus and framework for comparable taxonomic activities in other crop groups.

Collecting immature seeds

In the evolution of most arable crops there has been strong selection pressure for uniformity of seed ripening. This is an obvious benefit to farmers at harvest but also to collectors of genetic diversity, who have been able easily to acquire large samples of mature seed of landraces or old varieties from standing crops. In the case of wild relatives, and of forage grasses and legumes, seed ripening is usually spread over a period of time both within as well as between plants. In many grasses, for example, it is common to find that the mature seeds have been dispersed from the upper parts of the inflorescence while those at the base are still green and immature. This presents problems to the collector who, in these circumstances, will be able to collect some mature seed on the point of dispersal but who, in order to reach the desired sample size, may have to take seed which has not completed its normal maturation process. This seed will have to pass through its final developmental stages while abnormally separated from the mother plant. The important practical problem of maintaining seed in good condition during collecting expeditions has been reviewed by Smith (1985).

Since collectors in any one collecting season usually have many sites to visit and are unable to sit around waiting for seeds to ripen, the practical question is, how far away from full maturity can the collector safely go, when selecting seed? What will be the consequences to immediate viability, on the one hand, and to longevity on the other? The answers are likely to differ from one species to another, and will doubtless be influenced by the way the seed is handled between collection and subsequent drying in the genebank. It seems highly probable that the packaging and temporary storage of samples of immature seed of high moisture content, often at high ambient temperatures in the back of a vehicle or in the pannier of a mule, will seriously interfere with the normal processes of maturation.

Research is needed so that the magnitude of the problem and the means of dealing with it are understood in relation to the different species groups. Collectors need to be able to make the best compromises when attempting to apply optimum scientific procedures while faced with practical constraints in the field.

Not infrequently, it is either not possible to collect seed – due to immaturity of the seeds or to the grazing of inflorescences of forage species by wild or domestic animals, or undesirable to do so because the seeds are either too large – cocoa or coconut, or are difficult to maintain in a viable state until return to base – some tropical fruits. In these cases the new technique of collecting small pieces of tissue directly into *in-vitro* culture has been shown to be highly effective and it would be desirable to develop or adapt techniques so that the potential of the method could be exploited over a wider range of crops.

Genetic stability

We have already referred in previous chapters to the need for research projects on conservation topics such as:

Alternative low-cost, low-tech seed storage for developing countries in the humid tropics.

The potential of ultra-low moisture contents to compensate for higher storage temperatures.

Non-destructive viability testing.

Techniques for breaking seed dormancy prior to viability testing.

Methods of constructing and operating Core collections.

But there are, in addition, many other gaps in our knowledge in how best to conserve germplasm, some with applicability to conservation in general and others with relevance to particular species. For example, there is the worrying problem of genetic stability during artificial storage. It is known that seeds accumulate genetic changes as they age, and although storage at deep-freeze temperatures drastically slows biological processes, nevertheless the kind and frequency of genetic change in stored material needs to be more accurately assessed so that storage regimes may be modified, if possible, to minimise genetic change. This need extends also to storage *in vitro* which is known to be genetically hazardous; some kinds of *in-vitro* culture are purposefully used by plant breeders in order to induce genetic changes, the so-called somaclonal variation (Scowcroft, 1984). There is a need to quantify stability in different species under different storage regimes.

Cryopreservation is thought to bring cell activity to a complete stop and therefore to be a stable preservation method, but this remains to be verified experimentally.

The genetic costs of lowering storage standards

It has to be recognised that all too often, the standards which have been advocated for seed conservation – drying to 5 per cent moisture content in an atmosphere of 15 per cent relative humidity, and regenerating seed when the viability drops to 85 per cent – have been formulated by scientists from developed countries who no doubt had in mind the scale and standard of facilities to which they are accustomed. For many curators who are less well endowed with resources, these standards, however desirable, are at present unattainable, and it would be prudent to determine the consequences to the genetic integrity of the conserved germplasm, of adopting lower standards of storage and of regeneration. For example, seed drying in the humid tropics can be particularly difficult and seed multiplication is everywhere a resource-intensive business. Compromises are often necessary, but their costs should be understood, before rather than after they have been made.

Breeding systems of wild species

The maintenance of the genetic integrity of each accession during the seed multiplication process can be a major organisational challenge for curators, as discussed in Chapter 6. The methods adopted to prevent gene exchange by uncontrolled cross-pollination depend, to a large degree, on the breeding system of the species. Basic information on the breeding system is often lacking for wild species, particularly for tropical and sub-tropical species, and research needs to be done on this subject if accessions in these categories are to be adequately managed.

Quarantine is essential to the free movement of germplasm

It is a matter of extreme importance that, in the distribution of germplasm from genebanks, disease organisms are not inadvertently distributed with it. The numbers of possible pathogens of bacteria, fungi and viruses, to be found on or in different host species is alarmingly high. According to Hewitt (1977), *Phaseolus* beans, maize, potato and groundnut each have over 30 pathogens, wheat has over 70, peach 75, sugar cane and cotton 80 and apple over 150! The purposive spread of crop plants by man to new countries and continents has been accompanied, all too often, by the simultaneous but unwitting distribution of their pathogens.

The 1951 FAO Plant Protection Convention confined itself to setting out principles for ensuring that plants and parts of plants – including seeds – that are moved in the course of international trade, are healthy. It is left to national governments to draw up their own quarantine regulations in relation to whatever crops and their pathogens they think fit. Most countries are acutely aware of the possibilities and dangers of introducing new pathogens when trading in seed or other propagating material, and have introduced detailed and necessarily restrictive plant quarantine regulations for the protection of their agricultural and horticultural industries. These regulations can apply also to the exchange of plant genetic resources, which, when these consist of exotic crops or of wild species, of which categories the pathology tends to be less well understood then indigenous crop species, are often regarded as a greater potential danger.

As a result, the distribution of germplasm of the major crop groups world-wide is beset with a bewildering array of regulations, and the establishment of a database of all national requirements for the safe import of material of each crop or species, would greatly facilitate the movement of germplasm. It is imperative that quarantine regulations should be scrupulously observed and it is more likely that they will be observed, if curators are readily able to determine the requirements of each recipient country.

Regulations usually require certification of freedom from specified organisms, and hence detailed examination of the material is usually necessary. The method of examination depends on the pathogen whose presence is sought; in the simplest case it may be an examination of the seed, or plant tissue surface, for the presence of insect pests. Non-infested material can then be passed on to the consignee unharmed and unchanged. Alternatively, it may involve the raising of plants from a sample of seed and the examination of the plants for symptoms of the presence of a virus. Healthy plants, or seeds from them, can then be transferred to the consignee. Apart from the time delay inherent in this testing and multiplicating process, often at least one year, the selective multiplication of a small sub-sample inevitably distorts the genetic characteristics of the sample which was despatched by the curator of the donor genebank.

A third possibility involves the destructive examination of a sub-sample of the material, for the presence of fungi or bacteria within the seed. This procedure also reduces sample size and can affect genetic integrity. By no means all plant material is subject to quarantine examination and that which is, is subject to varying treatments according to species and country of origin. Therefore, generalisations are inappropriate, but the development of non-destructive tests for the detection of fungi, bacteria or viruses would undoubtedly be of great benefit. Some research has begun in this area, but more is needed.

Disease-resistance screens

Some of the best-known and most effective uses of genetic resources in the improvement of crop plants have been in the identification and transfer of genes for disease resistance – examples from potatoes and rice were given in Chapter 3, but the use of genetic diversity in breeding against the limiting effects of disease has occurred in many crops. In all cases the critical first step has been, not to screen the accessions for the presence of resistance alleles, but to develop a screening test which will permit this to be done.

The usual way to detect the presence of a disease-resistance allele is to allow it to express itself in the presence of the appropriate pathogen. Screening tests are, in effect, standardised procedures for challenging the plant with a particular pathogen under conditions which favour the development of an infection. Most diseases vary in their severity from year to year, because most are influenced in their expression by components of the environment; eelworms by soil moisture levels, rusts, mildews and moulds by atmospheric temperature and humidity, insect-vectored viruses by population density of the vector etc. Simply to grow the accessions to be screened in plots in the field, with the intention of observing the result, is rarely satisfactory. Absence of symptoms cannot be unambiguously attributed to the presence of a resistance allele; it may be due to the absence of inoculum or to unfavourable environmental conditions preventing the establishment of an infection.

In the case of resistances which are quantitative in their expression, 'horizontal resistances', and which are usually very subject in their expression to the modifying influence of environment, screening tests have a large component of error unless very carefully controlled. The purpose of the test is to indicate what degree of resistance would be exhibited if the material were to be grown in the field in a 'disease year', that is, in a year when environmental conditions favoured the pathogen.

For these reasons, screening tests are usually conducted under carefully defined conditions, often in glasshouse or laboratory, and in such a way that their results will reliably predict what would be the host–parasite interaction in the field. For some diseases, such as the cereal rusts, the inoculation and incubation of seedlings in a glasshouse is a relatively unsophisticated matter and the results may be obtained within a few days. For others, however, tests are technically more elaborate, requiring perhaps the use of growth cabinets and the inoculation of host material grown in closed containers, as in the case of quantitative resistance to potato cyst nematode derived from the wild potato species *Solanum verneii* (Phillips *et al.*, 1980).

Possibly the most complex resistance screening procedure is that described by P.H. Williams (1989), for the simultaneous challenging of a *Brassica* seedling with up to ten different pathogens, fungal, bacterial and viral, on roots, stem and leaves, as appropriate.

The development of a reliable screening test, which gives results which predict behaviour in the field, can itself be a considerable research project. Tests devised for this purpose will often be used subsequently as an essential part of breeding and selection programmes for the transfer of the resistance alleles which the screening has revealed. In each generation of the progeny under selection, a screening test must be employed to determine which individuals possess the critical resistance alleles, for it is only these individuals which are of use for further breeding. The effectiveness of a disease-resistance breeding programme is heavily dependent on the quality of the screening tests which it employs.

In conclusion, it should be stressed that the foregoing research projects are only a few selected examples by way of illustration. A comprehensive list of current and required research would be lengthy. It would also be impermanent; research needs change over time.

Defining a research programme for the support of crop genetic resources

In view of the diversity of topics and of crops on which research might profitably be done, of the nature of the research, which is essentially problem solving and which might not be taken up spontaneously by scientists in the appropriate disciplines, and of the limited funds which are available for research in genetic resources, there is clearly a need for a body with an oversight of the whole field to advise on the content of a programme and on its internal priorities.

Much of the funding for current research on crop genetic resources comes form the CGIAR and is channelled through the IBPGR, which has been mandated by its parent body to provide a lead to the genetic resources community in all aspects of genetic resources work. It seems sensible that IBPGR should continue to function as a stimulant and as a coordinating centre for genetic resources conservation research, but in its future operations it will doubtless pay close attention to advice from the expert membership of the international crop networks on what needs to be done. Some research needs will be crop specific, such as the development of resistance screening tests, the study of the gene-pool of the crop and its wild ancestors or the development of *in-vitro* culture protocols, and will have to be assessed against the competing claims of other crop-specific proposals. However, it may well be that essentially crop-specific proposals have relevance to needs in other crops, offering the possibility of aggregate projects. Other research needs will be more general, for example the development of low-cost storage methods and the development of the Core concept in genebank management. But whether specific or general, there is clearly a need for central strategic planning of research.

How can we ensure that the necessary research is done?

The conclusion to be drawn from the discussion above on research requirements, is the need for a coordinating body which, in one way or another, will see to it that the research is done in accordance with some agreed list of priorities. This role is filled at the moment by IBPGR which has two principal tasks. The first is to know, from a broad oversight of the world scene as well as from a detailed knowledge of the problems of national programmes and particular crop species groups, what needs to be done to strengthen and improve genetic resources conservation. The second is to know where, in the international scientific community, aspects of these needs are being met by ongoing research, or where it might be possible to catalyse scientists, working in adjacent areas, to do it at little or no additional cost. There will, of course, be a residual category of needs which are not being met, nor seem likely to be addressed in either of these ways. In these cases, the only sure way of getting this work done is to commission and pay for it.

The association of nations which constitute the CGIAR is currently providing some funds to IBPGR for commissioned research. The amounts, while inadequate, do represent a positive approach to this urgent need. The conventional way to organise research is to build, equip and staff a research institute to undertake the work. Such research institutes, are usually founded to work in particular areas of science, for an indefinite period of time and with a commitment to push back the frontiers of knowledge as far as they will go. But these are not the needs of plant genetic resources conservation, which require research to solve particular problems or to fill in particular gaps in the science base. When particular research objectives have been achieved, it will be necessary, immediately to redirect staff and resources to new problems. Flexibility, therefore, will be a very important requirement in the organisational arrangements for getting the research done.

The other important consideration is the range of scientific disciplines which is involved in the collection and conservation of plant genetic resources. These include agronomy, taxonomy, ecology, genetics, mycology, bacteriology and virology – the list is not exhaustive – applied by specialists to problems in, for example, species relationships, crop plant diversity, eco-geographical distributions of species, seed physiology, *in-vitro* studies, seed multiplication, and diverse aspects of plant pathology; again the list is not exhaustive.

To set up an institute competent to cope with this range of current activities and capable of responding to the needs of a changing programme, would be a huge and costly undertaking. It would also be unnecessary. There is a much more satisfactory alternative available in the contracting-out of research in the form of well-defined projects with a fixed budget and a finite time span to existing research groups which possess the highest levels of expertise in the appropriate scientific field. This method provides maximum flexibility, and gives access to the best laboratories and scientific expertise for as long as it is needed. Most importantly, it commits no funds to the construction of redundant monuments in bricks and mortar. This approach is being successfully pursued at the present time. Its only serious limitation can occur when it is not possible to find a laboratory with a sufficient level of available expertise to immediately start work on a problem, happily a rare event. In this case the only alternative seems to be to pay for a training phase, in which scientists from appropriate related disciplines acquire the necessary techniques. To some extent this difficulty has occurred in commissioning work on cryopreservation, where there are few centres of outstanding expertise. Otherwise, commissioned research has much to commend it as a way of strengthening the science base of germplasm conservation in the most cost-effective way.

The application of research results: technology transfer

The encouragement and commissioning of research as discussed above is motivated by the need to improve collection and conservation. It is a matter of great importance therefore that research results are applied in practice. Experience during the past fifteen years shows that, not infrequently, new knowledge is disregarded. We should enquire why.

The conventional method of communicating results of research is in the form of a paper to a learned journal, society or conference or as a monograph or report. All result in exposure of results to the scientific community; a process of passive information transfer.

A good example is the theory and practice of sequential viability testing by Ellis and Roberts (1984) (see Chapter 6). This work was published as part of a conference/symposium proceedings, in book form, and was widely distributed in the genetic resources community as well as being offered for sale in the usual way. Ellis and Roberts' idea, offered to curators the possibility of reducing the consumption of seed in viability testing and of reducing the routine work-load on genebank staff, without any increase in operating costs and without any requirement for new equipment. The application of the new technique is straightforward and the interpretation of results a simple matter. And yet, enquiries at meetings of curators reveal a general failure to adopt sequential viability testing in their genebank management routines.

At a recent rice germplasm workshop held at IRRI, more than half of those present claimed to be unaware of the content of a series of specialist IBPGR publications on long-term seed storage, which had been distributed free to all genebanks and other interested parties (Ellis, pers. comm., 1990).

These are two examples of what, regrettably, appears to be a common failure to get the results of research applied in practice. The nations which make up the CGIAR, expect the IBPGR to assume a leadership role in raising the standards of germplasm conservation. In stimulating research to fill gaps in the science base of conservation, IBPGR is discharging only part of that responsibility. It is clear that there is another part which should receive more attention than hitherto – active information transfer, the vigorous promotion of new ideas and techniques in collecting, conservation and genebank management. The greater use of workshops and hands-on training courses for genebank staff is one possibility, but an integral part of each research project should be a terminal phase when IBPGR ensures that all parts of the genetic resources community have the results and implications of the research brought to their attention and understanding. Simply to publish results in the conventional way and then to wait for things to happen, can be ineffectual and unacceptable.

8

Rising to the challenge

Résumé

In this chapter we place the issue of conservation of crop plant diversity within the wider context of the conservation of the biosphere and briefly discuss the principal issues on which progress depends, namely public awareness, organisation and money.

Present conservation projects are both numerous and diverse in their aims and are usually executed within the context of limited strategies.

Attempts have been made at coordinating strategies, recommendations proliferate but coordinated action is rare. There is a great need for an authoritative world conservation strategy and a UN agency with the resources to publicise its goals and to ensure their achievement. The conservation of crop plant diversity must be seen to be a special part of this wider approach to the protection of the biosphere.

Organisation of the scientific and technical inputs is also essential, with a greater use of multidisciplinary teams and a widening of objectives from the conservation of species to the conservation of ecosystems, within which we may also conserve the wild species relatives of crop plants.

The development and substitution of high-output sustainable systems of agricultural production for the current high-input high-output and traditional low-input low-output systems, will require new types of crop varieties capable of maximising yield without detriment to the environment and will increase dependence on the genetic control of diseases and pests. For both purposes the continued availability of genetic diversity will be an essential resource.

Finally, we end where we began in Chapter 1, by returning to the question of human population size and growth and its overriding influence on the health and stability of the biosphere.

The conservation of crop genetic diversity is part of biosphere protection

In this final chapter we briefly consider some of the issues affecting the future of conservation of crop genetic resources and try to place them within the wider context of the conservation of the biosphere.

As we have pointed out earlier, modern techniques of molecular biology already permit the transfer of genetic material between distantly, or indeed quite unrelated species which had hitherto been separated by barriers of absolute sexual sterility preventing gene exchange by conventional means. This possibility of side-stepping natural barriers, greatly extends the genetic resource base available to the breeder and focusses attention on the resources within wild species. For reasons discussed in Chapter 5, the conservation of wild species is best achieved by preservation *in situ* in their natural habitats, and this provides the direct link between the conservation of crop genetic diversity and the conservation of natural ecosystems.

Those concerned with the collection of crop genetic diversity are entitled to some feelings of satisfaction from their dramatic and effective rescue of much priceless crop diversity on the point of extinction. This notable achievement has occurred largely in isolation from other conservation activities on plant and animal species such as the establishment of nature reserves, the protection of endangered species of plants and particularly of large mammals, and of some plant communities which are of special scientific interest. All have been largely independent solutions to what were perceived as separate problems.

Because of the widespread and effective publicity given to a selected few of these individual conservation projects, there is a danger that the interested public is beginning to suffer from 'compassion-fatigue' and may wish to conclude that, on the whole, conservation problems including genetic conservation, are adequately taken care of. They are not. Despite the many hundreds, perhaps thousands of projects world-wide which are not widely publicised, the scale of the conservation effort is inadequate and from the point of view of the biosphere, fragmented.

There are three basic issues which have to be addressed if we are to come to grips with the problem of how we are to survive on the planet and yet ensure an adequate standard of living for our descendants. These are:

Defining and publicising the scale of the problems to be solved in biosphere protection.

Matching the scale of the inputs to the scale of the problems.

Organisational arrangements to ensure effective direction and use of resources.

Organisational issues

Political considerations

Much of what has been achieved to date for the world-wide conservation of germplasm of major crops, has been due to the efforts of national governments, non-governmental organisations, bilateral projects between donor governments and recipient countries and to the CGIAR, in which the IBPGR has played a central stimulating role. Within the field of crop genetic resources, there are some regional activities involving several countries and many species. An example is a new and ambitious project funded by the five Nordic countries: Denmark, Finland, Iceland, Norway and Sweden, operating in the ten countries who are members of the Southern African Development Coordination Conference, which aims to conserve diversity from a wide range of crop and forage species from the greater part of southern Africa. Other regional activities have different structures; for instance the countries of south Asia and of eastern Europe have regional consultative committees on genetic resources; others have founded an institute in one country to serve the needs of a group of neighbouring countries, for example, the Centro Agronomico Tropical de Investigacion y Ensenanza (CATIE), in Costa Rica.

Although differing widely in scope and objectives, all of these mechanisms and projects are similar in that they are attempts to deal with discrete parts of a larger whole, with necessarily limited objectives and usually with inadequate funding, especially over the long term.

Biosphere conservation, including aspects of the conservation of crop plant genetic diversity, is a world problem. National frontiers and national interests are of no significance to the distribution of species and to the boundaries of ecosystems, nor to the spread of aerial and marine pollution. Yet threatened ecosystems, in whole or in part, are under the control of sovereign states, and hence biosphere protection requires, on the one hand, the abandonment of narrow national interests and international political differences and the substitution of close international collaboration if it is to be effective and, on the other hand, a greater appreciation of the urgency of the problems, if action is to be taken in time.

While most of us would readily assent to the principle that we are faced with a world crisis requiring world-wide action, commonsense forces us to recognise the inequalities of opportunity which different countries have for responding to the crisis. Governments of too many developing countries, faced with appalling problems of poverty and malnutrition, of external debt burdens and of the increasing incidence of major disasters due to drought, flood and earthquake, turn first to the solution of immediate crises. These inevitably have first claim on their attention and on their already inadequate resources, even though they may well appreciate the magnitude of the long-term threat which they face from a degenerating environment. Indeed, no one can be more aware of the consequences of habitat destruction than the peasant farmer or pastoralist faced, in his or her daily struggle to survive, with an environment which is visibly deteriorating, partly because of his or her own activities. The perception of environmental degradation is likely therefore to be much more acute for the rural poor than for members of affluent industrialised societies, who may have a more general awareness of world problems, but from which they are at present cushioned, by their relative wealth. The greater the struggle to survive and the greater the perception of the need for change the less, it seems, is the opportunity for the individual or government to break out unaided and adopt more environmentally sensitive farming systems and industrial processes.

The only possible way out of this dilemma is for the developed world to recognise that its economic dominance carries with it the principal responsibility for saving the biosphere and for funding both the planning and execution of biosphere conservation on a scale far in excess of present efforts. We should not say that we cannot afford such responsibilities, for the bottom line is that we are part of the biosphere, we depend for our survival on it and on the genetic resources which it contains. For the sake of all our futures we must take care of it, not simply on moral grounds but out of self-interest.

Maximising the effectiveness of our efforts for biosphere protection clearly requires, first, intergovernmental cooperation, willing both the ends and the financial means to achieve them and, second, the mobilisation of trained scientists to execute a crash programme to deal with crises and to plan long-term strategies including the training of adequate numbers of young scientists, particularly in developing countries, competent to expand the work in the future.

How is the necessary degree of international collaboration to be achieved?

We lack precedents which might be used as a guide for achieving the degree and permanence of collaboration which is required within the context of an overall conservation strategy. A series of international organisations have been formed over the past twenty years with responsibilities for different aspects of conservation, which have achieved varying degrees of success.

The conservation issue was first placed on the international stage by the UN Conference on The Human Environment in 1972 (the Stockholm Conference), which gave rise to the United Nations Environment Programme (UNEP). Since then there have been seven major initiatives to raise the awareness of governments and taxpayers of the interconnectedness of environmental protection and economic growth (World Resources, 1986). These initiatives were:

The 1980 World Conservation Strategy commissioned by UNEP from IUCN.

The Inter-Parliamentary Union meeting in Nairobi, 1984.

The World Resources Institute's 'The Global Possible' Conference, 1984.

The Man and the Biosphere Programme of UNESCO, with which UNEP and IUCN are associated.

The establishment of the World Commission on Environment and Development in Geneva in 1983 and its report, *Our Common Future*, published in 1987.

The World Industry Conference on Environmental Management, Versailles, 1984.

The Global Meeting on Environment and Development for non-governmental organisations, convened in Nairobi in 1985 by UNEP.

In addition, there is an inter-agency conservation group, charged with coordinating strategies, on which FAO, UNESCO, UNEP and IUCN are represented with IBPGR present as an observer.

These high-level bodies have collectively produced hundreds of recommendations (World Resources, 1986), on what needs to be done and on the best ways of achieving it. The interest therefore is intense and the possibilities for confusion multiply with each successive set of recommendations. Two things are required if interest is to be translated into effective action.

The first requirement is for coordination of the policies and action plans of the different organisations. Moves to develop coordination began in 1980 when the World Conservation Strategy was released by IUCN, WWF, UNEP, FAO and others. A successor is being prepared for the 1992 UN Conference on Environment and Development. Whereas the original World Conservation Strategy stimulated the development of conservation strategies in more than forty-five countries, the new document will attempt to relate the major elements including, for example, genetic resources conservation and the creation of nature reserve areas, to topics which have been more prominent in the last decade, especially economic aspects, population growth, the harnessing of the traditional knowledge and practices of indigenous peoples and ethical issues concerning the ownership and exploitation of genetic diversity. It seems to us important that this revised document should be sufficiently authoritative for all nations and conservation organisations to be able to unite behind its principles and to use it as an overall guide to concerted action.

The other prime requirement is for one organisation, perhaps an expanded UNEP, to be given by common consent, a leadership role to stimulate and guide the practical realisation of the aims of the World Strategy. Whatever body is either created or expanded to fill this need, it must have adequate resources to fulfil its responsibilities. We are looking for more than assent to a set of principles to guide national policies and actions. These are necessary, but must be preceded by an overall strategy and followed by detailed and continuous activity.

This body should have three principal tasks:

Defining objectively the problems to be addressed in conserving the biosphere, without regard for the vested interests of particular organisations or countries.

Producing strategies for addressing the problems it had defined, based on up-to-date science as well as political and economic realities, placing existing conservation efforts within this strategic framework and commissioning new projects to fill gaps in accordance with agreed priorities.

Mobilising and channelling resources, financial and human, on a scale commensurate with the defined needs.

It could provide a focus at the highest level through which all countries could formally commit themselves to support in cash or in kind and through which developed countries could direct funds, expertise and training to support the needs of an international collaborative conservation programme. In the case of the crop genetic resources component, this should also include breeding activities to utilise the genetic diversity conserved for the improvement of crop plants. Conservation for its own sake is insufficient. Both highly bred and traditional crops must change in a changing world and genetic diversity has an essential and continuing contribution to make.

There are hopeful signs that the unprecedented degree of cooperation which is required is likely to materialise. On the one hand, there is the recent lessening of hostility between political and ideological power blocs and the diversion of resources from armaments to other fields, and, on the other, an increasing public awareness that mankind may be facing a greater threat to well-being from the degradation of the environment than from hostile economic or military forces. Some of the threats have been well publicised, such as the increase of atmospheric carbon dioxide, sulphur dioxide and chlorofluorocarbons, the destruction of rain-forests and the spread of deserts in Africa. Others, such as the conservation of crop genetic resources, are scarcely understood by the general public. It is important the conservation of crop genetic resources should become part of the common understanding, and be seen as a vital component of the wider conservation movement, which is essential to our ability to fit our crops of the future to the needs of a changing world.

Scientific effort needs organising too

There is a growing realisation that future action should be directed towards the conservation of ecosystems rather than to selected components of these communities and should be part of a more general need for a holistic approach to the conservation of the biosphere. Attention has too often been focussed on those species which are aesthetically or emotionally appealing, or are perceived to be of immediate or possible future economic value. The idea that other constituent species of ecosystems are also worthy of conservation is now gaining acceptance, not because these species might be of economic value in the future – though some undoubtedly will be – but because the constituent species of an ecosystem are mutually adapted and need each other if they are to survive and if the system is to continue to function as an integrated whole. The conservation of the wild species relatives of our crop plants is but a part, although a special part, of the larger task of ecosystem preservation.

One of the features of the development of biological science has been the reductionist approach to its organisation. Areas of science in which an individual can be fully competent have become narrower as knowledge has increased in depth. Biology has fragmented into many discrete areas of enquiry which have developed with a large degree of independence from each other. This progressive focussing of intellectual effort into analytical study has led to tremendous advances in our understanding of how living things function and reproduce, culminating in the recent revelations of gene structure and function.

However, for the solution of many of the practical problems encountered in the conservation of crop genetic diversity and even more so in the conservation of ecosystems, this traditional method of organising science must be supplemented or replaced by a collaborative approach, where specialists are brought together in multidisciplinary teams which have the range of scientific and technical expertise to tackle the complex problems encountered. Many models exist of how this may be done and of the benefits which can accrue. For example, the success of many large plant-breeding projects at the International Agricultural Research Centers can be attributed to the collaborative work of specialists in genetics, agronomy, virology, mycology, ecology, nematology, taxonomy, economics and sociology. The same method of working has been successfully applied to some of the more recent aid and development schemes. Team working, which requires some limitation of conventional freedom of enquiry for the common good, is becoming more common and may well become the normal way of organising projects in the future.

If the urgent aims of a World Strategy are to be achieved, nothing less than a crash programme in the mobilising and deployment of scientific personnel will do. In many disciplines, experienced scientists are now scarce – the paucity of taxonomists has already been referred to. Wherever suitable people can be found they should be induced to turn their attention to conservation problems, providing both leadership in project planning and execution and accepting a major training role to increase the size of the pool of scientists, competent to implement the global strategy.

At the level of the individual, the magnitude and diversity of the problems which are emerging in the conservation of crop genetic resources, present exciting opportunities for young scientists looking for fields in which to employ their skills. The scientific and organisational problems are formidable and they offer a worthwhile and challenging occupation. While in many developed countries the opportunities for doing conventional biological and agricultural research are shrinking rapidly, the conservation of genetic diversity and its use in crop improvement in developing countries offer alternative routes to career development and personal fulfilment in working for the benefit of others.

Food production and the environment

Two major issues which appear at the moment to be fundamentally opposed to each other are the desperate need to increase food production in the poorest countries and the equally urgent requirement to modify food production systems so that they become 'sustainable'.

Sustainable development is development that meets the needs of the present without compromising the ability of future generations to meet their own needs (*Our Common Future*, World Commission on Environment and Development, 1987). The reader is referred to this work for a full discussion of the many aspects of sustainability, economic, industrial, agricultural and social, and their implications for the realisation of human aspirations. Here, we wish to draw attention to the close relationship between agricultural sustainability and crop genetic resources.

It is now widely accepted that increased food production must come from the land already under cultivation and not from the expansion of agriculture into areas still under natural ecosystems. Farming has spread as far as it usefully can. What remains is, in general, too hostile an environment for food production and indeed much traditional farming land has been degraded by overgrazing, overcultivation or excessive irrigation. We must therefore resolve the conflict between the need to extract more food from a declining farming area and the need to do it in a sustainable way.

The idea of sustainable cropping and livestock production has, for many people, the implication of lower-input, more 'natural', ecologically sensitive systems with, inevitably, lower productivity. There is a movement in some developed countries towards 'alternative' agriculture which is environment-friendly and extols what are thought to be the simple virtues of past traditions. This is not the same thing as sustainable agriculture which has as its principal aim the maximising of production without damage to the environment. This will undoubtedly call for novel rather than traditional methods of farming.

The high-yielding varieties which dominate crop production at present have mostly been developed for systems with high energy inputs in the form of artificial fertilisers and pesticides, usually with mechanisation, and often with irrigation water and sophisticated management practices. The possibilities of maintaining output, after the substitution of new low-input systems, are currently under investigation and the resolution of this problem is a matter for conjecture. But it is certain that if the rural poor are to be helped into new sustainable ways of improving their standard of living and if the developed world is to maintain that to which it has become accustomed, new types of crop varieties and management systems will be required. Control of diseases and pests by genetic means will undoubtedly be given much higher priority in future, both for high and low-input systems. For both of these objectives – sustainability and genetic control of disease – the availability and use of crop plant genetic diversity is essential and the conservation of what nature has provided in the past for use in the future is a matter of supreme importance.

A hopeful sign, is the increasing awareness in the taxpaying public of the developed countries, especially among the young, that the planet does not exist for the thoughtless exploitation by man, but is a tolerant but ultimately destructible biosphere to which we have a responsibility for care. The signs are being read and interpreted that the living world is humanity's life support system, but there is some way to go before a sufficient weight of public opinion can be brought to bear on governments to recognise their responsibilities to posterity, and to take action on a scale and with the permanence which is required.

In summary, we emphasise the interconnectedness of genetic diversity, crop improvement, food supplies, nutritional levels, poverty, habitat destruction and the overriding driving force, the rocketing rise in world population, particularly in the developing countries.

Population control

One of the themes which we have tried to stress throughout this book is the close interconnectedness of all the components of the biosphere and we have discussed some of these connections in detail; plants with plants, plants with man, and plants with man and the physical environment in which we have our being. The cardinal issue, on which we have not elaborated since we drew attention to its overriding importance in Chapter 1, but which has intimate connections with all the other factors, is the explosive increase in human populations – one billion more bodies to feed and house in the present decade alone. It has been said that man has done more damage to the biosphere in the last fifty years than in the previous 5000 and this threatening cloud hangs over all attempts to conserve and save the biosphere and to ensure a reasonable standard of living for all peoples.

The issue of population control is outside the scope of this book, but we cannot bring the book to a conclusion without recognising that swift and effective control measures are essential if we are to succeed in conserving the environment and crop genetic diversity. This is both the ultimate challenge and the primary task and is the most important example of the inescapable interconnection of human activity with the health of the biosphere.

References

Austin, R.B., Bingham, J., Blackwell, R.R., Evans, L.T., Ford, M.A., Morgan, C.L. and Taylor, M. (1980). Genetic improvements in winter wheat yields since 1900 and associated physiological changes. *Journal of Agricultural Science, Cambridge*, **94**: 675–689.

Bettencourt, E. and Konopka, J. (1990). Directory of Germplasm Collections. *4. Vegetables.* IBPGR, Rome, 174–175.

Bradshaw, A.D. (1975). Population structure and the effects of isolation and selection. In *Crop Genetic Resources for Today and Tomorrow*, pp. 37–51. Frankel, O.H. and Hawkes, J.G. (eds). Cambridge University Press, Cambridge.

Breese, E.L. (1989). *Regeneration and Multiplication of Germplasm Resources in Seed Genebanks: The Scientific Background.* IBPGR, Rome, 1–69.

Brown, A.D.H. (1989). The case for core collections. In *The Use of Plant Genetic Resources*, pp. 136–156. Brown, A.D.H., Frankel, O.H., Marshall, D.R. and Williams, J.T. (eds). Cambridge University Press, Cambridge.

Burton, W.G. (1966). *The Potato.* Veenman and Zonen, Wageningen.

Campbell, G.K.G. (1976). Sugar beet, *Beta vulgaris* (*Chenopodiaceae*). In *Evolution of Crop Plants*, pp. 25–28. Simmonds, N.W. (ed.). Longman, London.

Carson, R. (1962). *Silent Spring.* G. Fawcett Publications, Greenwich, CT.

Chang, T.T. (1976). Rice, *Oryza sativa* and *O. glaberrima* (*Gramineae–Oryzeae*). In *Evolution of Crop Plants*, pp. 98–104. Simmonds, N.W. (ed.). Longman, London.

Chang, T.T. (1989). The case for large collections. In *The Use of Plant Genetic Resources*, pp. 123–135. Brown, A.D.H., Frankel, O.H., Marshall, D.R. and Williams, J.T. (eds). Cambridge University Press, Cambridge.

Clark, C. (1967). *Population Growth and Land Use.* Macmillan, London.

Ellenby, C. (1952). Resistance to the Potato Root Eelworm, *Heterodera rostochiensis* (Wollenweber). *Nature*, **170**: 1016.

Ellis R.H. and Roberts, E.H. (1984). Procedures for monitoring the viability of accessions during storage. In *Crop Genetic Resources; Conservation and Evaluation*, pp. 63–76. Holden, J.H.W. and Williams, J.T. (eds). George Allen and Unwin, London.

Ellis, R.H., Hong, T.D. and Roberts, E.H. (1985). *Handbook of Seed Technology for Genebanks.* Vols 1 and 2. IBPGR, Rome.

FAO (1984). *FAO Production Yearbook, 1983*, Vol. 37. FAO, Rome.

Feldman, M. (1976). Wheats, *Triticum* spp. (Gramineae–Triticinae). In *Evolution of Crop Plants*, pp. 120–128. Simmonds, N.W. (ed.). Longman, London.

Ford-Lloyd, B.V. and Williams, J.T. (1975). A revision of the genus *Beta* sect. *vulgaris* and new light on the origin of cultivated beet. *Journal of the Linnean Society, Botany*, **71**: 89–102.

Frankel, O.H. (1984). Genetic perspectives of germplasm conservation. In *Genetic Manipulation: Impact on Man and Society*, pp. 161–170. Arber, W., Llimensee, K., Peacock, W.J. and Starlinger, P. (eds). Cambridge University Press, Cambridge.

Frankel, O.H. and Bennett, E. (1970). *Genetic Resources in Plants – Their Exploration and Conservation*. International Biological Programme, Handbook No. 11. Blackwell Scientific Publications, Oxford.

Gale, J.S. and Lawrence, M.J. (1984). The decay of variability. In *Crop Genetic Resources: Conservation and Evaluation*, pp. 77–101. Holden, J.H.W. and Williams, J.T. (eds). George Allen and Unwin, London.

Gale, M.D. and Youssefian, S. (1985). Dwarfing genes in wheat. In *Progress in Plant Breeding – 1*, pp. 1–35. Russel, G.E. (ed.). Butterworth, London.

Gregory, R.S. (1987). Triticale breeding. In *Wheat Breeding; its Scientific Basis*, pp. 269–286. Lupton, F.G.H. (ed.). Chapman and Hall, London.

Grove, R. (1990). Threatened islands, threatened earth; early professional science and the historical origins of global environmental concerns. In *Sustaining Earth*, pp. 15–29. Angell, D.J.R., Comer, J.D. and Wilkinson, M.L.N. (eds). Macmillan, London.

Harlan, J.R. (1971). Agricultural origins: centers and non-centers, *Science, NY*, **174**: 468–474.

Harlan, J.R. (1976). The plants and animals that nourish man. *Scientific American*, **235**: 89–97.

Harrington, J.F. (1970). Seed and pollen storage for conservation of plant gene resources. In *Genetic Resources in Plants – Their Exploration And Conservation*. International Biological Programme, Handbook No. 11, pp. 501–522. Frankel, O.H. and Bennett, E. (eds). Blackwell Scientific Publications, Oxford.

Hawkes, J.G. (1983). *The Diversity of Crop Plants*. Harvard University Press, Cambridge, MA.

Hawkes, J.G. (1990). N.I. Vavilov – the man and his work. *Biological Journal of the Linnean Society*, **39**, No. 1: 3–6.

Hewitt, W.B. (1977). Plant health problems of a general nature. In *Plant Health and Quarantine in International Transfer of Genetic Resources*, pp. 3–16. Hewitt, W.B. and Chiarappa, L. (eds). CRC Press, Cleveland, Ohio.

Holdgate, M. (1990). Changes in perception. In *Sustaining Earth*, pp. 79–96. Angell, D.J.R., Comer, D.J. and Wilkinson, M.L.N. (eds). Macmillan, London.

IBPGR (1988). *Descriptors for Papaya*. IBPGR, Rome.

IBPGR (1989). *Annual Report for 1988*. IBPGR, Rome.

IBPGR (1990). *Annual Report for 1989*, IBPGR, Rome.

Iwanaga, M. (1987). Use of wild germplasm for sweet-potato breeding. In *Exploration, Maintenance and Utilization of Sweet-Potato Genetic Resources*, Report of the First Sweet-Potato Planning Conference, pp. 199–210. International Potato Center (CIP), Lima, Peru.

Larter, E.N. (1976). Triticale, *Triticosecale* spp. (Gramineae–Triticinae). In

Evolution of Crop Plants, pp. 117–120. Simmonds, N.W. (ed). Longman, London.

Lupton, F.G.H. (1987). History of wheat breeding. In *Wheat Breeding: Its Scientific Basis*, pp. 51–70. Lupton, F.G.H. (ed.). Chapman and Hall, London.

MacNeill, J. (1990). Sustainable development. Meeting the growth imperative for the 21st century. In *Sustaining Earth*, pp. 191–205. Angell, D.J.R., Comer, D.J. and Wilkinson, M.L.N. (eds). Macmillan, London.

Marshall, D.R. and Brown, A.H.D. (1975). Optimum sampling strategies in genetic conservation. In *Crop Genetic Resources for Today and Tomorrow*, pp. 53–80. Frankel, O.H. and Hawkes, J.G. (eds). Cambridge University Press, Cambridge.

Odum, S. (1965). Germination of ancient seeds. Floristical observations and experiments with archaeologically dated soil samples. *Dansk Botanisk Arkiv*, **24**: 1–70.

Phillips, L.L. (1976). Cotton, *Gossypium (Malvaceae)*. In *Evolution of Crop Plants*, pp. 196–200. Simmonds, N.W. (ed.). Longman, London.

Phillips, M.S., Forrest, J.M.S. and Farrer, L.A. (1980). Screening for resistance to potato-cyst nematode using closed containers. *Annals of Applied Biology*, **96**: 317–322.

Plucknett, D.L., Smith, N.J.H., Williams, J.T. and Murthi-Anishetti, N. (1987). *Gene Banks and the World's Food*. Princeton University Press, Princeton, New Jersey.

Purseglove, J.W. (1982a). *Tropical Crops. Dicotyledons*. Ipomoea L. pp. 78–88. Longman, London.

Purseglove, J.W. (1982b). *Tropical Crops. Dicotyledons*. Hevea, pp. 144–171. Longman, London.

Purseglove, J.W. (1982c). *Tropical Crops. Dicotyledons*. Gossypium, pp. 333–364. Longman, London.

Purseglove, J.W. (1985a). *Tropical Crops. Monocotyledons*. Oryza L. pp. 161–199. Longman, London.

Purseglove, J.W. (1985b). *Tropical Crops. Monocotyledons*. Musaceae, pp. 343–384. Longman, London.

Purseglove, J.W. (1985c). *Tropical Crops. Monocotyledons*. Dioscorea, pp. 97–117. Longman, London.

Roberts, E.H. (1975). Problems of long-term storage of seed and pollen for genetic resources conservation. In *Crop Genetic Resources for Today and Tomorrow*, pp. 269–295. Frankel, O.H. and Hawkes, J.G. (eds). Cambridge University Press, Cambridge.

Roberts, E.H., King, M.W. and Ellis, R.H. (1984). Recalcitrant seeds; their recognition and storage. In *Crop Genetic Resources: Conservation and Evaluation*, pp. 38–52. Holden, J.H.W. and Williams, J.T. (eds). George Allen and Unwin, London.

Roberts, H.F. (1919). The founders of the art of breeding – 2. *Journal of Heredity*, **10**: 147–152.

Scowcroft, W.R. (1984). *Genetic Variability in Tissue Culture; Impact on Germplasm Conservation and Utilisation*. IBPGR Special Report, IBPGR, Rome.

Simmonds, N.W. (1976). Bananas, *Musa* (Musaceae). In *Evolution of Crop Plants*, pp. 211–215. Simmonds, N.W. (ed.). Longman, London.

Smith, R.D. (1985). Maintaining viability during collecting expeditions. In

Report of Third Meeting of IBPGR Advisory Committee on Seed Storage, Appendix 3, pp. 13–22. IBPGR, Rome.

Toxopeus, H.J. and Huijsman, C.A. (1952). Genotypical background of resistance to *Heterodera rostochiensis* in *Solanum tuberosum* var. *andigenum*, *Nature*, **170**: 1016.

Wickens, G.E., Haq, N. and Day, P. (1989). *New Crops for Food and Industry*. Chapman and Hall, London.

Williams, J.T. (1989). Plant germplasm preservation; a global perspective. In *Biotic Diversity and Germplasm Preservation, Global Imperatives*, pp. 81–96. Knutson, L. and Stoner, A.K. (eds). Kluwer Academic Publishers, Dordrecht.

Williams, P.H. (1989). Screening for resistance to diseases. In *The Use of Plant Genetic Resources*, pp. 335–352. Brown, A.D.H., Frankel, O.H., Marshall, D.R. and Williams, J.T. (eds). Cambridge University Press, Cambridge.

Withers, L.A. (1984). Germplasm conservation *in vitro*: present state of research and application. In *Crop Genetic Resources: Conservation and Evaluation*, pp. 138–157. Holden, J.H.W. and Williams, J.T. (eds). George Allen and Unwin, London.

Withers, L.A. (1989). *In vitro* conservation and germplasm utilisation. In *The Use of Plant Genetic Resources*, pp. 309–334. Brown, A.D.H., Frankel, O.H., Marshall, D.R. and Williams, J.T. (eds). Cambridge University Press, Cambridge.

World Commission on Environment and Development (1987). *Our Common Future*. Oxford University Press, Oxford.

World Resources Institute and International Institute for Environment and Development (1986). *World Resources 1986*. Basic Books Inc., New York.

Wycherley, P.R. (1976). Rubber, *Hevea brasiliensis* (Euphorbiaceae). In *Evolution of Crop Plants*, pp. 77–81. Simmonds, N.W. (ed.). Longman, London.

Zeven, A.C. and de Wet, J.M.J. (1982). *Dictionary of Cultivated Plants and Their Regions of Diversity*, pp. 21–31. Centre for Agricultural Publishing and Documentation, Wageningen.

Index

157